Synthesis Lectures on Information Concepts, Retrieval, and Services

Editor
Gary Marchionini, *University of North Carolina, Chapel Hill*

Synthesis Lectures on Information Concepts, Retrieval, and Services publishes 50- to 100-page publications on topics pertaining to information science and applications of technology to information discovery, production,distribution, and management. For more information, visit
`www.morganclaypool.com/r/icr`.

Knowledge Management (KM) Processes in Organizations: Theoretical Foundations and Practice
Claire R. McInerney and Michael E. D. Koenig
2011

Search-Based Applications: At the Confluence of Search and Database Technologies
Gregory Grefenstette and Laura Wilber
2010

Information Concepts: From Books to Cyberspace Identities
Gary Marchionini
2010

Estimating the Query Difficulty for Information Retrieval
David Carmel and Elad Yom-Tov
2010

iRODS Primer: Integrated Rule-Oriented Data System
Arcot Rajasekar, Reagan Moore, Chien-Yi Hou, Christopher A. Lee, Richard Marciano, Antoine de Torcy, Michael Wan, Wayne Schroeder, Sheau-Yen Chen, Lucas Gilbert, Paul Tooby, and Bing Zhu
2010

Collaborative Web Search: Who, What, Where, When, and Why
Meredith Ringel Morris and Jaime Teevan
2009

Knowledge Management (KM) Processes in Organizations

Theoretical Foundations and Practice

Copyright © 2011 by Morgan & Claypool

Knowledge Management (KM) Processes in Organizations: Theoretical Foundations and Practice

Claire R. McInerney and Michael E. D. Koenig

www.morganclaypool.com

ISBN: 9781598299564   paperback
ISBN: 9781598299571   ebook

DOI 10.2200/S00323ED1V01Y201012ICR018

A Publication in the Morgan & Claypool Publishers series
SYNTHESIS LECTURES ON INFORMATION CONCEPTS, RETRIEVAL, AND SERVICES

Lecture #18
Series Editor: Gary Marchionini, *University of North Carolina, Chapel Hill*
Series ISSN
Synthesis Lectures on Information Concepts, Retrieval, and Services
Print 1947-945X   Electronic 1947-9468

Knowledge Management (KM) Processes in Organizations

Theoretical Foundations and Practice

Claire R. McInerney
Rutgers, The State University of New Jersey

Michael E. D. Koenig
Long Island University, NY

SYNTHESIS LECTURES ON INFORMATION CONCEPTS, RETRIEVAL, AND SERVICES #18

 MORGAN & CLAYPOOL PUBLISHERS

ABSTRACT

Knowledge Management (KM) is an effort to increase useful knowledge in the organization. It is a natural outgrowth of late twentieth century movements to make organizational management and operations more effective, of higher quality, and more responsive to constituents in a rapidly changing global environment. This document traces the evolution of KM in organizations, summarizing the most influential research and literature in the field. It also presents an overview of selected common and current practices in knowledge management, including the relationship between knowledge management and decision making, with the intention of making a case for KM as a series of processes and not necessarily a manipulation of things. The final section highlights the use of social networking and commonly adopted Web applications to increase the value of social capital and to connect practitioners with clients and colleagues.

KEYWORDS

knowledge management, information management, knowledge sharing, content management, CMS, Content Management Systems, communities of practice, corporate Intelligence, organizational intelligence, organizational learning, or the learning organization

Contents

Acknowledgments

The authors gratefully acknowledge the contributions made to this work by William Voon, who gave valuable assistance with the research, and Scott Young without whose very able assistance the construction of this book would have been far more onerous.

Claire R. McInerney and Michael E. D. Koenig
December 2010

CHAPTER 1

Introduction

1.1 WHAT IS KM?

There are numerous definitions of KM, quite literally scores of them. Indeed, Professor Michael Sutton [2008] of the Gore School of Business at Westminster College reported at the ICKM (International Conference on Knowledge Management) meeting in 2008 that he had assembled a library of more than 100 of them.

Three classic definitions of KM ones are presented here. At the very beginning of the KM movement, Davenport, T. (1994) offered the following:

> "knowledge management is the process of capturing, distributing, and effectively using knowledge"

This definition has the virtue of being simple, stark, and to the point.

A few years later, the Gartner Group created the second definition of KM, which is perhaps the most frequently cited one (Duhon, 1998):

> A discipline that promotes an integrated approach to identifying, capturing, evaluating, retrieving, and sharing all of an enterprise's information assets. These assets may include databases, documents, policies, procedures, and previously uncaptured expertise and experience in individual workers

This definition is a bit more specific and informative, and it is illuminating because it makes explicit not just conventional information and knowledge units, but also "tacit knowledge," or implicit knowledge, that which is known, but not captured in any formal or explicit fashion.

The third definition by McInerney, C. [2002] is that "KM is an effort to increase useful knowledge within the organization. Ways to do this include encouraging communication, offering opportunities to learn, and promoting the sharing of appropriate knowledge objects or artifacts." This definition emphasizes the interactive aspect of KM, that is, knowledge sharing by people rather than the common understanding of knowledge management as a system used to organize what we might call 'knowledge objects.' These objects or artifacts are means of expression that represent knowledge, and they are usually stored in digital form on some kind of enterprise resource system or on a Web-based platform such as an Intranet or wiki. Knowledge objects can enable the open interchange of useful knowledge that has been codified, but there are many other forms of knowledge interaction that take place in low tech environments or merely among individuals working together.

Upon being asked at a cocktail party to define Knowledge Management, one may offer an apt definition, comprised of primarily 3 distinct parts:

1) Classic Library and Information Science and Information Retrieval.

2) ICT, Information and Communication Technology.

3) HR, Human Relations, changing the culture of the organization to facilitate knowledge sharing and use.

Another way to understand KM may be to examine its history and development.

1.2   THE HISTORY AND DEVELOPMENT OF KM

The appearance of the term "Knowledge Management" is a rather recent phenomenon. It appeared operationally only in the mid-1990s. The earliest print reference appears to have been used first in the context of library and information work. Marchand, D. [1985], then Dean of the School of Information Studies at Syracuse University, coined it in the 1980s as a descriptor for the final level in his stage hypothesis of information systems development [Koenig, M., 1992a]. However, the term, as presently used, appears to have been re-coined more or less anonymously somewhere among the major accountancy and consulting firms. The earliest reports seem to be from McKinsey & Co. Brook Manville, the first Director of Knowledge Management at McKinsey, reports that McKinsey launched an internal study whose title included the phrase Knowledge Management in 1987 (Manville, personal communication to MEDK, 2007). Larry Prusak, one of the acknowledged pioneers of the field, in a very corroborative communication, reports that Ernst & Young started using the term in 1992, and that McKinsey & Co. was using it "a bit earlier – about 1988?" (Prusak, personal communication to MEDK, 2007). Note that the early definition of KM cited above by Davenport (1994) stems from his work at Ernst & Young.

In the article entitled "Where Did Knowledge Management Come From" written in 1999, Prusak states that KM first appeared "about seven years ago," i.e., in 1992, and describes a conference in Boston in 1993 as "a good milestone to mark the beginning of the knowledge management time-line" [Prusak, L., 1999]. The conference was held at the Four Seasons Hotel in the spring of 1993 and was sponsored by the Center for Business Strategy of Ernst & Young (Prusak, personal communication to MEDK, 2007).

The earliest instances of KM, as the term is understood today, derive from the consulting world, from which the principles of KM eventually spread to other disciplines. The consulting firms quickly realized the potential of the Intranet flavor of the Internet for linking together their own geographically dispersed knowledge based organizations. They then understood that the expertise they had gained was a product that could be sold to other organizations. That product needed a name, and the name chosen, or at least arrived at, was Knowledge Management. In his article discussing the origins of KM, Prusak in fact, felt it appropriate to say "some skeptics may believe that consultants developed knowledge management to replace declining revenues from the waning re-engineering movement" [Prusak, L., 1999]. The two enthusiasms, KM and re-engineering, are related in that both were driven by increased ICT (Information and Communication Technologies) capabilities. However, the timing, though convenient for the consulting firms, was driven not by their convenience, but by the straightforward dynamics of ICT capability growth. KM was dependent upon the appearance of the Internet, while pre internet, ICT already enabled the major restructuring of an organization's work flows and processes. In a sense, KM also has roots in the implementation of Supply Chain Management (SCM) software and business process reengineering (BPR) as well as the more recent development of Enterprise Resource Planning (ERP). IT development has always displayed a pattern, of growth from more structured data to less tractable, less well structured, or comparatively unstructured data. Remember that text processing was at one time called "string handling," because to people brought up on handling numeric data, text was most conveniently thought of as a string of symbols. In that sense, SCM & BPR & ERP to KM, represent a logical and predictable progression toward unstructured information and knowledge.

Another aspect of KM's relationship to ICT is that KM emerged at approximately the same time as the cost of personal computers dropped to the degree that PC's became cost effective and affordable desktop tools for the ordinary person. PC adoption made it clear that the ease of personal computing was going to allow individuals to readily share what they knew and what they wrote about through file transfer and email. Collections of data and information files stored in shared files and attached to email messages could be moved without the risk of losing manila folders filled with important paper reports. Even early "if-then" projections on Visicalc (the pioneer spreadsheet program) were an enormous boost to managers and entrepreneurs who required flexible and malleable financial data to inform decisions. Even though spreadsheets deal with data, it is the ability to manipulate data and learn from them and form decisions that must also be considered as one of the rudimentary foundations of knowledge management.

Today there are numerous voices critical of the name "Knowledge Management." These objections arise from a perceived inadequacy of the term Knowledge Management, which detractors say lacks strong description while conflicting with the concept that one cannot inherently manage knowledge. "The Nonsense of Knowledge Management" (Wilson, T. [2002]) is the best of the genre. The argument that one cannot manage knowledge is, however, not very persuasive; after all, Webster's Collegiate Dictionary (10th edition) has nine flavors of meaning for 'manage' and eight for knowledge', and a number of those 72 possible combinations do indeed make good sense. The argument that KM is a very poor descriptive term has much more merit. The replacement phrase most often suggested is "Knowledge Sharing." One of the more recent substitutions for the term 'Knowledge Management' is 'knowledge flow management,' preferred and used by Leistner, F. [2010] in his book *Mastering Organizational Knowledge Flow*. Leistner, Chief Knowledge Officer in SAS' Global Professional Services' division, has a background in computing and experience with IBM's KM Institute. He is deeply conversant with the social issues inherent in knowledge sharing and the way knowledge is developed. Although he claims that everyone makes up his or her own definition of knowledge management, Leistner says that knowledge must be connected with people to be a viable term.

Such substitutions, however well merited, are unlikely ever to be made. The situation with KM is quite analogous to the very similar concern a few decades back that "word processing" was a lousy descriptor. One processed meat, of course, but surely not words. Words made poetry. However, word processing was the term the industry, then principally IBM and Wang, chose to use, and that is the term that stuck. Now we take the term word processing for granted, seldom giving it a second thought, and only old timers remember the vitriol with which it was attacked. The adoption of a new term to replace "knowledge management" is no more likely than it was for "word processing."

One of the more recent substitutions for the term 'Knowledge Management' is 'knowledge flow management,' preferred and used by Leistner, F. [2010] in his book *Mastering Organizational Knowledge Flow*. Leistner, Chief Knowledge Officer in SAS' Global Professional Services' division, has a background in computing and experience with IBM's KM Institute. He is deeply conversant with the social issues inherent in knowledge sharing and the way knowledge is developed. Although he claims that everyone makes up his or her own definition of knowledge management, Leistner says that knowledge must be connected with people to be a viable term.

Given this background information, another good functional definition of KM and how it developed is the equestrian metaphor of "by the intranet out of intellectual capital" [Koenig, M., 2000a]. By this definition, KM has two parents, the enthusiasm for and the appreciation of intellectual capital, and the development of the Internet and its offspring, intranets and extranets. Intellectual Capital is, in turn, a token of the larger recognition of the importance of information and knowledge. "Intellectual Capital" representing the awareness that as Peter Drucker [Hibbard, J., 1997, p. 46] put it:

> We now know that the source of wealth is something, specifically human knowledge. If we apply knowledge to tasks that we obviously know how to do, we call it productivity. If we apply knowledge to tasks that are new and different, we call it innovation. Only knowledge allows us to achieve those two goals.

Indeed, Drucker is clearly one of the intellectual fathers of KM. His *Post Capitalist Society*, which appeared in 1993, is about the shift from industrial capitalism to the knowledge society. Almost one full page of an 11 page index is devoted to knowledge or knowledge phrases such as knowledge workers, yet the phrase knowledge management is surprisingly absent [Drucker, P., 1993]. The early use of the phrase 'Knowledge Management' is sometimes attributed to Drucker, but at this point, it appears to be still unsubstantiated. Given Drucker's stature and his emphasis on knowledge and knowledge workers, it is almost inevitable that such an attribution would arise. It may be noted, however, that while Drucker did, carefully, claim a first for the phrase Knowledge Worker, he does not appear to have made any similar claim for the phrase Knowledge Management, and he was certainly active and involved well into the KM time frame in which he likely would have made such a claim if the thought it warranted.

Tom Peters was an early user of and promoter of the phrase Knowledge Management, or as it was known then Knowledge Management Structure (KMS). In his 1992 book *Liberation Management* [Peters, T., 1992], he has four chapters specifically named Knowledge Management

Structures (I – IV: Taking Knowledge Management Seriously; Getting Physical; Knowledge Bases, Expert Systems, Computer Augmented Collaboration; and Developing and Tapping Expert Power in the Hierarchy-less Organization), almost 10% of a very large book. Although the KMS phrase did not take hold, Peters was clearly writing about KM as it came to be understood.

INTELLECTUAL CAPITAL DEVELOPS AND DECLINES

A brief review of the development of Intellectual Capital (covered in detail elsewhere: [Koenig, M., 1996, 1998, Sullivan, P., 2000] is vital to understanding the origins and development of KM.

A pioneer in the Intellectual Capital (IC) field was Karl-Eric Sveiby [Sveiby, K., 1989, 1997, 2001], whose book *The Invisible Balance Sheet* was a key work in the development of thinking about Intellectual Capital. From this early writing, Sveiby's ideas developed smoothly and rapidly into KM. The work of Buckman, R. [2004] at Buckman Laboratories and Hubert St. Onge [Chatzkel, J., 2000] at the Canadian Imperial Bank of Commerce and Clarica Insurance were key implementations in operationalizing the concept and in popularizing IC. Also very important was the Stewart, T. [1994] article in *Fortune* magazine, "Intellectual Capital, Your Company's Most Valuable Asset."

The first blush of enthusiasm for Intellectual Capital centered on quantifying and measuring it [Edvinsson, L., 1994, Edvinsson and Malone, 1997b]. Measuring such an ill-defined commodity as intellectual capital, however, proved to be difficult. Clearly, if it was so important, it needed to be measured. However, as the difficulty of measuring such an amorphous commodity as information, much less knowledge, became apparent to the business community, the enthusiasm for intellectual capital decreased, as evidenced by the reduced volume of related publications. The best metrics system to use for quantifying the "amount" of knowledge shared or the value of KM in the organization are still key questions for researchers who study the field.

THE INTERNET EMERGES

As the Internet emerged, the business world realized that the Internet could be used to link an organization together. This was the take off point for large scale recognition of KM as an important innovation, and it was the stimulus for its development [Koenig, M., 1996, 1998].

1.3   THE STAGES OF KM DEVELOPMENT

In observing the development of KM as practiced, described, and discussed at professional meetings, conferences, and trade shows, one can observe three clear stages.

STAGE ONE

The initial stage of KM was driven primarily by information technology, or IT. Organizations, particularly the large international consulting organizations, realized that their stock in trade was information and knowledge.

These groups also realized that internal communication and information sharing was often lacking. If knowledge could be shared more effectively, then the efficiency would increase business and the bottom line would improve. When the internet emerged, they realized that the intranet flavor of the internet provided a valuable tool to accomplish knowledge coordination and sharing. The first stage of KM focused on the deployment of new technology to accomplish these information sharing goals.

Those large international consulting organizations, then realized quickly that many of their customers shared exactly the same problems, and that the expertise they were building for themselves could also function as a product, an expertise that they could purvey to those customers. A new product needs a name and a theme or rationale. The name for their new product was Knowledge Management. The crucial thematic justification for KM was *intellectual capital*, a theme that had emerged as a burgeoning topic in the business literature just a few of years earlier. To revisit a much quoted aphorism, the first stage might be described as the "If only Texas Instruments knew what Texas Instruments knew" stage [Davenport and Prusak, 1998a, O'Dell and Jackson, 1998]. The hallmark phrase of Stage 1 was first "***best practices***" to be replaced by the more politic "***lessons learned***," all dependent to some extent on a technological framework.

STAGE TWO

The second stage of KM can be described simply as adding the recognition of the importance of the human and cultural dimensions. The second stage might be described as the, 'if you build it they will come' is a fallacy stage. In other words, the recognition that building KM systems alone is not sufficient and can easily lead to quick and embarrassing failure if human factors are not sufficiently taken into account. As this recognition unfolded, two major themes from the business literature were brought into the KM fold. The first was Senge's work on the learning organization in his seminal work, *The Fifth Discipline: The Art and Practice of the Learning Organization* [Senge, P., 1990]. The second was Nonaka and Takeuchi's [1995] work on tacit knowledge and how to discover and cultivate it, *The Knowledge-Creating Company: How Japanese Companies Create the Dynamics of Innovation*. Both were not only about the human factors of KM implementation and use, they were also about knowledge creation as well as knowledge sharing and communication. The hallmark phrase of Stage 2 was "***communities of practice***," an important concept taken up later in Stage 3 by Etienne Wenger, Richard McDermott, and William M. Snyder (2002) in their key work *Cultivating Communities of Practice: A Guide to Managing Knowledge*.

STAGE THREE

The third stage was the awareness of the importance of content, and, in particular, an awareness of the importance of the retrievability and, therefore, of the importance of the arrangement, description, and structure of that content. Since a good alternate description for the second stage of KM is the "it's no good if they don't use it" stage, then in that vein, perhaps the best description for the new third stage is the "it's no good if they can't find it" stage, or perhaps "it's no good if they try to use it, but can't find it." Another bellwether is that TFPL (a major U.K. information and knowledge management consultancy), in their report of the October 2001 CKO (Chief Knowledge Officer) Summit that they hosted, reported that for the first time taxonomies emerged as a topic, and it emerged full blown as a major topic. The hallmark phrases emerging for the third stage are *content management* (or enterprise content management) and *taxonomies*. In 2006, a two-day long "Taxonomy Boot Camp" was added to the KMWorld conference. KMWorld is both a journal and a website devoted to Knowledge Management concepts, practices, and events.

NOW, STAGE FOUR?

Is KM entering a new growth stage, a fourth stage, or is it merely maturing?

Perhaps a fourth stage of KM may be emerging, although the delineation of this fourth stage is not nearly as clear as for the previous stages. This stage can be characterized by the awareness of the importance of information and knowledge external to the organization.

The inclusion of information and knowledge external to the organization is not new to KM, one need only think of the World Bank and the high visibility of their KM program and their very deliberate development of communities of practice (CoPs) specifically designed to encompass experts external to the bank, to make that point [Srikantaiah, T., 2004]. That being said, however, the overwhelming emphasis of KM to date has been to mobilize and make accessible the organization's information and knowledge and to rely on technology to make that happen. Perhaps the most basic mantra of KM has been the "if only we knew what we knew," the classic metaphoric example of what KM is all about, i.e., making *an organization's* knowledge more useable and more productive. If one returns to the definitions of KM at the beginning of this lecture, one can clearly see that organizational emphasis.

Several threads have converged to drive this new emphasis:

The extension of Intranet based KM systems to Extranet based systems.

The first and most obvious application of Internet technology was to make an "intranet" of it, to use the Internet as an access controlled network for the company. A next logical step was to use the same access control mechanisms to build an "extranet" so that persons outside the organization, vendors, suppliers, dealers, major customers, etc., could be included. This has proceeded much more slowly, since security, particularly for for-profit corporations is a major concern (discussed later in this book). How do you know that your competitor is not posing as a vendor of your products, or has not acquired a vendor of your products precisely so as to have access to some of your information? It was this lack of concern about competitive information that allowed the World Bank to so quickly move into extranet based communities of practice [Srikantaiah, T., 2004]. However, as security techniques have improved, the willingness of corporations to extend communities of practice onto extranets so as to incorporate the knowledge of "outsiders" has correspondingly increased. A good example is Caterpillar Inc. [Stuedemann, R., 2007], who brought in dealers and independent repair facilities to their communities of practice. They agonized for months over what sort of non disclosure form to use to allow access from outside the organization, produced what they jokingly called 'the mother of all non-disclosure forms,' and then discovered that those outside the firm barely glanced at it before signing [Stuedemann, R., 2004].

Concern about the soon to be lost knowledge held by post-war baby-boomers as they begin to hit retirement age. Corporations are beginning to realize that KM, in the form of communities of practice, is a wonderful tool to address this lost knowledge issue. If retirees are encouraged to remain as active members of one or more communities of practice, then their knowledge is not lost, and the retiree may well be able to contribute valuable knowledge (Delong, D. [2004]). This is discussed in Chapter 7.

A repeat of the same broadening phenomenon that occurred with MIS, Management Information Systems.

After Management Information Systems (MIS) were introduced in the 1970s, there came a rebound of disillusionment with the field of MIS. Much of that disillusionment was a function of the fact that what the MIS system contained was typically only the organization's raw and partially aggregated transaction information, purchases, production data, sales, etc., useful data and information to be sure, but not the information constituting the bulk of what an executive typically needed for those decisions near the top of the managerial decision making pyramid. For those decisions, what was typically needed was external contextual data, not the transactional data in the typical MIS system. This realization drove an awareness of the importance of external information, and it also drove a partial rebranding of MIS, re-launched as DSS, Decision Support Systems, a terminology designed to avoid overselling those transactional data based systems and to avoid implying that they had all the data or information that an executive needed. That same awareness is now developing in the KM world, the awareness that much of the key information needed for critical decisions lies outside the corporation or organization, and that the ideal KM system should provide appropriate links to the world outside the organization.

The result is a greatly increased emphasis upon external information. It should also be noted that another emphasis emerging simultaneously is that of the importance of situating information and knowledge in context. This is, in fact, another facet of the same evolution, the awareness that the importance, the usability, and the value of information is a function of how it relates to other information. New knowledge derives from the combination of information, either the juxtaposition of existing information, or the addition of new information to existing knowledge. If I have seen further than others, it is because I have stood on the shoulders of giants.

Is this really a new Stage 4? Or just a maturation of KM? Probably the best answer is the latter. The distinction is hardly so clear as those between the previous stages, and the continuity is greater, with context emphasis long apparent in many cases. Yet it is a very heightened emphasis, and that emphasis is an important theme of this book.

Summary of the Stages of KM Development

The Three Stages of KM

STAGE I "By the Internet out of Intellectual Capital"

Information Technology

Intellectual Capital

The Internet (including intranets, extranets, etc.)

Key Phrases: "***best practices***," later replaced by the more politic "***lessons learned***"

STAGE II Human and cultural dimensions, the HR, Human Relations stage

Communities of Practice

Organizational Culture

The Learning Organization (Senge), and

Tacit Knowledge (Nonaka) incorporated into KM

Key Phrase: "***communities of practice***"

STAGE III Content and Retrievability

Structuring content and assigning descriptors (index terms)

Key Phrases: " ***content management***" and "***taxonomies***"

STAGE IV ? Access to External Information

Emphases upon External Information and the recognition of the Importance of Context

Key Terms: "***context***" and "***extranet***"

1.4 SUPPLEMENTARY WAYS OF LOOKING AT KM

1.4.1 THE IBM TWO BY TWO MATRIX
A MAP OF THE DOMAIN OF KNOWLEDGE MANAGEMENT

KM may also be displayed and to a degree defined graphically through mapping. The following presents an expanded form of a graphic used by IBM in their KM consultancy to explain the value and purpose of KM.

DOMAINS OF KNOWLEDGE MANAGEMENT STRATEGY		
	COLLECTING (STUFF) & /CODIFICATION/	CONNECTING (PEOPLE) & PERSONALIZATION
DIRECTED INFORMATION & KNOWLEDGE SEARCH	Databases, external & internal	community & learning
	Content Architecture	directories, "yellow pages" (expertise locators)
	Information Service Support (training required)	findings & facilitationg tools, groupware
EXPLOIT	data mining best practices / lessons learned/after action analysis (HARVEST)	response teams (HARNESS)
SERENDIPITY & BROWSING	Cultural support current awarness profiles and databases selection of items for alerting purposes / push	Cultural support spaces - libraries & lounges (literal & virtual), cultural support, groupware
EXPLORE	data mining best practices (HUNTING)	travel & meeting attendance (HYPOTHESIZE)

(Note the dashed lines are intended to indicate that the boundaries are porous and overlapping. One can only admire the clever mnemonics of the Cs, Es, and Hs.)

From: Tom Short, Senior consultant
Knowledge Management
IBM Global Services

1.4.2   THE FOREST AND THE TREES

KM - The New Business Potpourri or not Seeing the Forest for the Trees

The late 20th century was full of business enthusiasms, hot topics, and fads, but KM, as demonstrated in Chapter 2, has become quite different from any of the rest in terms of its sustained growth and staying power. It is quite literally unique among those topics and enthusiasms. The argument that we make here is that it is not just quantitatively different, but that it is also quite qualitatively different, and different in a fundamentally important way.

What is that fundamentally important difference?

Think of all the management fads and enthusiasms of the late 20th century, 1975 – 2000. What is striking is how many of those management fads, enthusiasms, and topics are highly related with the management of information, knowledge flow in organizations or the management of information technology (IT).

Below is a list of those management fads, enthusiasms, and topics that meet those criteria. There have been quite a few, indeed. There are surprisingly few other hot business enthusiasms that do not meet those criteria.

Information / Knowledge related business enthusiasms and hot topics of the last quarter century

(Listed in approximate chronological order with the most recent first; note that these are topics, not specific dated events)

Enterprise Content Management (ECM)

Supply Chain Management (SCM)

Customer Relationship Management (CRM)

Enterprise Resource Planning (ERP)

Knowledge Management (KM)

Intellectual Capital (IC)

E-business

Data Warehousing / Data Mining

Core Competencies

Business Process Re-Engineering

The shift from Hierarchies to Markets, both economic and political

Competitive Intelligence (CI)

Total Quality Management (TQM) and Benchmarking

Information Technology (IT) and Organizational Structure

Information Resource Management (IRM)

Enterprise-Wide Information Analysis (IBM Inc.)

Management Information Systems (MIS) to (Decision Support Systems (DSS) and the importance of External Information

I.T. as Competitive Advantage

Managing the Archipelago (of Information Services)

Information Systems Stage Hypotheses (Nolan, Rockart, Gibson & Jackson, Marchand, Koenig, &Zachman)

Decision Analysis

Data Driven Systems Design (the fundamental basis of Structured Programming)

I.T. and Productivity

Minimization of Unallocated Cost

(Note, there is not sufficient room in this text to elaborate on these topics, but they are defined and discussed in some detail in Koenig, M. [2000a].)

The conclusion that one can draw is that the topics above are the trees in a forest, a forest of information and knowledge (small 'k') management, whose scope and importance we are still coming to recognize. Furthermore, it is beginning to appear that KM is graduating from being just one of many names on that list, to now becoming the name for that forest of all the trees of information and knowledge (small 'k') management.

What is also striking is that after a quarter of a century of business fads, there have been no new significant business trends in the last few years. The conventional explanation offered for that paucity of new movements is that the dot-com bust created a period of skepticism and a climate unreceptive to new enthusiasms and new topics, yet with the prompt return of a boom economy, at least until 2008, one would have expected new fads to emerge, but they did not.

The more likely explanation is that of the forests and the trees, the forest being that community of trees listed above that all deal significantly with information and knowledge management. The forest is certainly not going away, nor will it remain static, new trees will emerge, but KM is morphing and expanding in scope to be the name of that forest. We have always had trouble defining KM, and now we have another definition, or more exactly a new metaphor, **KM is the name for that newly recognized forest of all the trees of information and knowledge (small 'k') management.**

1.4.3 KM AS THE EXTENSION OF THE SUCCESSFUL R&D ENVIRONMENT

Meta-Research, or KM is the Extension of the Successful R&D Environment
A final way to view KM is to observe KM as the movement to replicate the information environment known to be conducive to successful R&D - **rich, deep, and open communication and information access** - and deploy it broadly across the firm.

The principles and practices of KM have developed in a very conducive environment, given that in this post-industrial information age, an increasingly larger proportion of the population consists of information workers. The role of the researcher, considered the quintessential information worker, has been studied in depth with a focus on identifying environmental aspects that lead to successful research [Allen, T., 1977, Goldhar et al., 1976, Koenig, M., 1990, 1992a, 2005, Mondschein, L., 1990, Orpen, C., 1985]. It is a logical development then to attempt to apply those successful environmental aspects to knowledge workers at large. The embarrassing fact, embarrassing at least to those who conducted research on the research environment, is that there is little evidence that this body of knowledge had any direct impact on the development of KM, a classic example of the silo phenomenon. KM appears to have developed in the business community quite independently of any knowledge of the body of research literature about research effectiveness. Nevertheless, this definition of KM as the extension of the successful R&D environment may ultimately prove to be the most straightforward and the most illuminating. An aside is that "Silo," as in a container/building with no windows, is used frequently in the KM literature to refer to bits of an organization that do not interact with the rest of the organization.

CHAPTER 2

Background Bibliographic Analysis

One measure of the influence of a discipline is to track the "formal communications" or published works in that discipline [Koenig, M., 2005, Ponzi, L., 2004]. Ponzi observed that "knowledge management is one emerging discipline that remains strong and does not appear to be fading"[Ponzi, L., 2004, p. 9]. Articles about KM were and are being published in the fields of computer science, information systems, management, engineering, communication, and library and information science. Ponzi's research on knowledge management publications is deep and comprehensive, but limited in that his latest results are from 2001. Ponzi and Koenig [2002] were able to project early on that KM was either an unusually broad-shouldered business enthusiasm or a rather permanent development. The authors have continued that tracking of the KM literature time series (Figure 2.1 below) through the 2009 literature. The KM business literature continues to grow.

Note that Figure 2.1 almost certainly underestimates the size of the KM literature. In the early years of KM, it was probably a very safe assumption that almost all KM articles would have the phrase "knowledge management" in the title, but as the KM field has grown, that almost certainly is no longer a safe assumption. There are now numerous articles about "communities of practice" or "enterprise content management" or "lessons learned" that clearly are KM focused, but they do not use the phrase "knowledge management" in the title.

The significance of the KM growth pattern becomes much more apparent when one compares it with the pattern of other major business enthusiasms of recent years. Below (Figure 2.2) are the literature growth patterns of three of those major business enthusiasms. The difference is dramatic. Quality Circles, Business Process Engineering, and Total Quality Management all show an almost identical pattern of approximately five years of dramatic, exponential, growth, then they peak and fall off to near nothing almost as quickly. KM, by contrast, has that same period of five years of exponential growth, 1994 to 1999, but in the decade since it has not declined, rather it has continued to grow steadily and consistently. All the hallmarks are here of a rather permanent development.

There has also been substantial interest in the academic world concerning KM. The database 'Dissertations and Theses' includes bibliographic information about theses published by graduate students at accredited North American institutions from 1861, and from 50 European universities since 1988. A search of the database showed that all of the dissertations and theses with 'knowledge management' in the title or in the key word fields have been published since 1996. The specific departments and disciplines in which the dissertations were written range from mathematics to mass

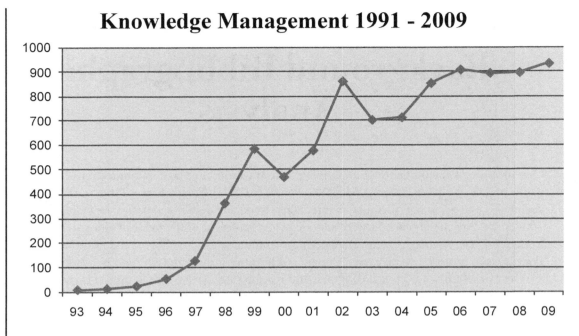

Figure 2.1: Knowledge Management Growth. Number of KM articles published by year.

communication, with business administration being strongly represented. See Figure 2.3 below for the publication pattern. In general, the number of dissertations focusing on some aspect of knowledge management rises gradually until 2006 and has remained steady with about 100 theses produced each year in English with, however, a decline in 2008 and 2009.

The number of scholarly papers and dissertations devoted to KM demonstrates that there has been and continues to be a scholarly interest in knowledge management even if that research has taken a small downturn. An examination of the types of research being conducted shows that over the years the subject matter of KM studies has changed somewhat from an emphasis on technological systems to a focus on communication and interaction among people. Also, the terminology has changed as well. Although 'knowledge management' may have been commonly used in the late 1990's and early 2000's, scholars have adopted terms such as 'knowledge sharing,' 'communities of practice,' and 'learning organizations' as knowledge management processes became more mainstream in organizations. As the twenty-first century has progressed, searches on 'knowledge management' have revealed that scholarly works on knowledge sharing have increasingly been combined with research on social networking and social media.

The data seem to indicate that there continues to be a lively interest in research and writing about knowledge management, and presumably that scholars and ordinary people are interested

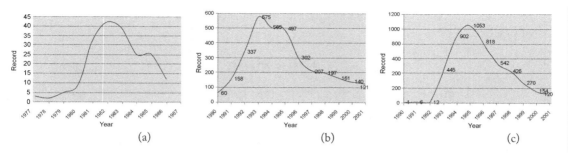

(a)   (b)   (c)

Figure 2.2: Literature graphs for "Quality Circles," "Total Quality Management," and "Business Process Reengineering."
(a) based on Abrahamson, 1996; (b) based on Ponzi and Koenig, 2002; (c) based on Ponzi and Koenig, 2002.

in reading about KM as well. The specific departments and disciplines in which the dissertations were written range from mathematics to mass communication, with business administration being strongly represented. See Figure 2.3 for the publication pattern.

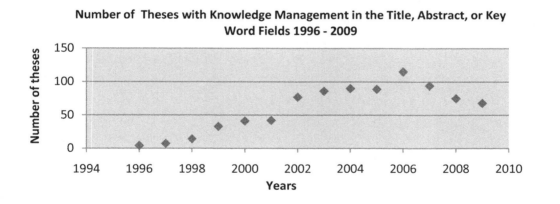

Figure 2.3: Doctoral Dissertations and Masters Theses written with 'Knowledge Management' in the Title, Abstract or Key Word Fields 1996–2009.

An interesting observation is that there was a very brief spurt of articles about KM in journals devoted to education, but that interest soon waned. This is likely a function of the fact that KM, as mentioned previously has a very corporatist and organizational emphasis, while for most academic principals, the faculty, their commitment to their field, their discipline and sub-discipline, their "invisible college" comes first. Their commitment to their nominal home institution is quite secondary. And, for most of those faculty, their invisible college already functions as their community of practice.

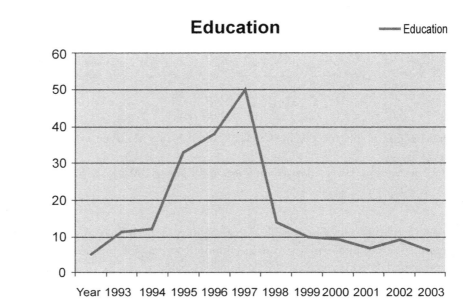

Figure 2.4: The Number of KM Articles in the Education Literature.

CHAPTER 3

Theorizing Knowledge in Organizations

This chapter provides an overview of the development of research findings and theories related to knowledge management.

In order to better understand the notion of "managing" knowledge, there is a need to better understand what it is about knowledge flow in organizations that lends itself to any form of management. The literature has discussed organizational knowledge both as a resource [Grant, R., 1996] and a process of learning [Argyris and Schon, 1978, Senge, P., 1990], often emphasizing one aspect over the other. In the resource view, knowledge is conceptualized as an object that exists largely in formal documents or online artifacts amenable to organizing and manipulation. The process view, on the other hand, largely emphasizes the emergent nature of knowledge that is often embedded within a person or within organizational routines, activities, and outcomes, or arises from the interplay of persons and existing information or knowledge. While both perspectives may vary significantly in terms of the scope for the "management" of knowledge, it is still worth exploring the issues and debate surrounding the practice of creating, gathering, and sharing knowledge within organizations.

3.1 KNOWLEDGE AS RESOURCE AND PROCESS

Through the resource perspective, organizations view knowledge as a fundamental resource in addition to the traditional resources of land, labor, and capital. It is held that the knowledge that the firm possesses is a source of sustainable competitive advantage, and is, accordingly, regarded as a strategic resource of the firm in need of management attention. On the other hand, through the process view, organizations are thought of as information processing and knowledge generating systems [Grant, R., 1996]. In the course of innovation and production of goods and services, information and knowledge are regarded as central inputs to organizational processes. Learning and knowledge are then seen as direct outcomes of activities performed commensurate with the organization's central mission and core competencies. Whether as a resource or as a process, for organizations that have begun to recognize organizational knowledge as a source of competitive advantage, knowledge generation and retention have become strategic necessities for such knowledge dependent firms.

Baumard, P. [1999] proposes looking at knowledge in organizations along two dimensions: tacit-explicit versus individual-collective. He defines four quadrants in which knowledge types are situated: tacit-individual (intuitiveness), tacit-collective (social practice), explicit-individual (expertise), and, explicit-collective (rules). Grounding the use of the quadrants in observations of exemplar case-study organizations, Baumard suggests that the creation of organizational knowledge can be tracked by locating actors' responses (knowing) within the appropriate quadrants of the matrix.

3.2   INTERACTIONS FOR KNOWLEDGE CREATION

While knowledge itself may be perceived as a resource, its creation occurs through human interactions, whether physical or virtual. For example, for knowledge to emerge from within a group, interactions that occur among its members shape the knowledge that emerges from the mutual engagement and participation of the group members. Those with a communication and interaction perspective have argued that through discourse and dialectics, individuals shape and re-shape the thought processes of others, eventually leading to a situation of negotiated or mutually co-constructed reasoning for action and knowledge [von Krogh et al., 1998]. Sense-making [Weick, K., 1995] is then seen as an activity that reaffirms whether the decisions and actions taken are rational in hindsight, constituting the "knowledge" that is created. Nonaka and Takeuchi [1995] in their seminal work have also alluded to knowledge creation as a process of socialization that is predicated on the need for direct social interactions.

Nonaka and Takeuchi are the most prominent theorists in the knowledge management domain. Their SECI (Socialization, Externalization, Combination, Internalization) model posits a spiral-type process in which knowledge goes from within a person's own knowledge store to a more explicit state that can be shared socially with others. This happens through a series of transformations that involve externalization and combination of what a person learns with experience and beliefs and then to the internalization stage where one takes what is learned and incorporates it within. If viewed through these ebbs and flows of thought, that is, socialization, externalization, combination, and internalization, knowledge creation takes on a very dynamic character, always changing, always synthesizing. In Nonaka's (1994) well-known paper on "A Dynamic Theory of Organizational Knowledge Creation," he argues in detail and through case studies that the processes related to knowing are continually transformative [Nonaka, I., 1994].

3.3   ACTIVITY AS CONTEXT

Instead of examining knowledge *per se*, Blackler, F. [1995] and others propose that attention should focus on systems through which knowing and doing are achieved. By suggesting an alternative stance of knowing as mediated, situated, provisional, pragmatic, and contested, as opposed to a more classic view of knowledge as embodied, embrained, encultured, and encoded, Blackler recognizes that knowledge permeates activity systems within the organization. Building on Engeström, Y. [1999] general model of socially distributed activity systems, Blackler, F. [1995] proposes that knowledge can be observed as emerging out of the tensions that arise within an organization's activity systems, that is, among individuals and their communities, their environment (rules and regulations), and the instruments and resources that mediate their activities. Through immersion in joint activity, individuals in organizations gain tacit knowledge, the sharing of which occurs as a result of the mutual participation [Tsoukas, H., 1996].

CHAPTER 4

Conceptualizing Knowledge Emergence

4.1 GATEKEEPERS, INFORMATION, STARS, AND BOUNDARY SPANNERS

A substantial body of research has been developed on the transmission of information within organizations, particularly R&D organizations.

The seminal work was that of Thomas J. Allen of MIT [Allen and Cohen, 1969, Allen, T., 1977] who conducted a number of studies relating to information flow in industrial and corporate R&D laboratories. Allen's most ingenious contribution to the field was to seize upon the phenomenon that in many cases in the context of military R&D and procurement, the same contract is awarded to two different organizations to achieve the same end, typically in the case of a critical component of a larger system. Duplicative development contracts may, in fact, be very worthwhile insurance against the failure of a key component of a system. This duplication provided a wonderfully robust context in which to examine information flows and what distinguished the information flows in the more successful projects from the less successful.

Allen coined the term 'Gatekeeper' to describe the information flow stars that he discovered, the heavily connected nodes in the information flow pattern. The reason that he chose that term was that much of the development and project work that he investigated was classified military work, where there seemed to be something of a paradox, how was a team to be successful if it didn't effectively connect with the world of information outside the organization? But how did it do that in a classified and communication restricted environment? What he discovered was that the information stars, the sociometric stars, were the answer to that paradox; they were the information channels through which external information reached the project team. That role was so crucial in the contexts that Allen typically investigated what he termed his sociometric stars "Gatekeepers." They oversaw and guarded the gates through which external information reached the projects. Indeed, one might say that they were not just the gatekeepers, they themselves were the gates.

The terminology is understandable, given that context, but a bit misleading just the same, and rather too narrow, for the gatekeepers did much more. They were also the channels for information sharing and exchange within the organization and within the project. Allen himself, in fact, in developing and explicating the role of gatekeepers introduces and explains his gatekeepers with the term "sociometric stars." "Information stars" a term emerging later [Tushman and Scanlan, 1981a,b], is, however, a more apt description, one that brings to mind more of the multiple roles and functions that such persons perform.

Allen found that the more productive teams were particularly characterized by having had more diverse information contacts outside the project team than did the less productive teams. In particular, he elucidated and illuminated the rich informal communication networks, typically quite independent of the formal organizational structure, characteristic of the ore successful companies, and management's relative unawareness of either he importance of, and in many cases, even the existence of those networks.

Furthermore, the "information stars" were central to information flow both within the organization at large, and within their project or projects. The characteristics that distinguished these stars were:

extensive communication with their field outside of the organization

greater perusal of information sources, journals, etc., information mavens

a high degree of connectedness with other information stars, one can infer that their utility was not just having more information at their fingertips, but knowing to whom to turn within the organization for further information

an above average degree of formal education compared to their project teammates

These characteristics of information stars were further corroborated by Mondschein, L. [1990] in a study of R&D activities across several industries.

One of the more intriguing of Allen's findings, given the context and the nature of security restrictions, was that the more successful teams made less use of external consultants. At first glance, this seems surprising, because wouldn't external consultants help bring in that external information that would be helpful? The answer seems to be that the more successful teams had better gatekeepers. The teams with better gatekeepers needed external consultants less, and consequently, used them less.

Another finding was that the information flow structure was not at all closely related to the formal organizational structure, and that the information stars did not map onto any consistent pattern of organizational placement or level. The relationship between formal organizational structure and the information flow structure also seems to be in part a function of the larger corporate culture. For example, Frost and Whitley [1971] adopted Allen's techniques to examine information flow in R&D labs in the U.K., and they found a somewhat higher overlap between formal organizational structure and the information flow structure than Allen had found in the U.S. There is a suggestion here that the more rigid the organizational hierarchy, the more the information flow structure is constrained to adapt itself to the formal organizational structure.

Tushman, M. [1977], Tushman and Scanlan [1981a,b] further extended the Allen tradition. Tushman examined development activities, both at the departmental level and at the project level, at a medical instruments company, and very much confirmed Allen's conclusions. He introduced and added the concept of "boundary spanning" or boundary spanner to describe very much the same phenomenon that Allen described as gatekeeping. He extended Allen's work by distinguishing between two types of communication stars, "internal communication stars" and "external communication stars," and defining boundary spanners as those who were both internal and external communication stars. The emphasis is clearly directed to projects and project management, and the "take home" theme is that boundary spanners should be recognized, utilized, and nurtured for facilitating project success.

In the context of KM, this tradition relates very directly to the development of **Communities of Practice** (CoP). Given the relative non-alignment of organizational structure and information flow and sharing, CoPs can be seen as the setting up of an alternative structure to facilitate information flow and sharing.

4.2   RESEARCH PRODUCTIVITY AND KNOWLEDGE

The 'Gatekeepers, Information Stars & Boundary Spanner' tradition is very consistent with a substantial body of work studying research productivity. Koenig, M. [1992a], for example, in the context of the U.S. pharmaceutical industry, studied the relationship between research productivity and the information environment in which that research was conducted. The productivity measure was, at base, simply the number of approved new drugs (new drug applications or NDAs) per millions of dollars of R&D budget. This measure, however, was refined by weighting the NDAs in regard to: 1) whether or not the Food and Drug Administration (FDA) judged the drug to be an "important therapeutic advance," 2) the chemical novelty of the drug, and 3) the filing company's patent position in regard to the drug, an indicator of where the bulk of the research was done. The study is compelling because of the high face validity of the measure of success, the successful introduction of new pharmaceutical agents, since that is what pharmaceutical companies are about after all, and because of the statistical robustness of the results, a consequence of the fact that the more successful companies were found to be not just twenty or thirty percent more productive than the not so successful companies, they were two or three hundred percent more productive.

The more productive companies were characterized by:

A relatively egalitarian managerial structure with unobtrusive status indicators in the R&D environment,

Less concern with protecting proprietary information,

Greater openness to outside information, greater use of their libraries and information centers, specifically, greater attendance by employees at professional meetings,

Greater information systems development effort,

Greater end-user use of information systems and more encouragement of browsing and serendipity. Increased time spent browsing and keeping abreast,

Greater technical and subject sophistication of the information services staff.

Note the relationship with research/project success and a corporate culture that is relatively egalitarian and enjoys relatively unobtrusive status indicators. Also most interesting is that the correlation with concern for protecting the confidentiality of proprietary information was negative, and it was also the strongest single correlation with research project success, and with an unassailably high statistical significance.

This topic is well covered in a recent encyclopedia article [Koenig, M., 2009] "Productivity Impacts of Libraries and Information Services," in *the Encyclopedia of Library and Information Sciences*: 3rd Edition.

There are also, in this overall tradition, two books in recent years that have been very well received and that offer valuable insights about information flow and information use in a modern organizational environment, Davenport and Prusak [1998a] *Working Knowledge*, and Brown and Duguid [2000a] *The Social Life of Information*.

4.3   LACK OF RECOGNITION OF THESE FINDINGS IN THE BUSINESS COMMUNITY

As Allen pointed out in his study, there is a surprising lack of recognition of these findings about the importance of information stars in the business community. This is, in fact, a subset of an even larger problem - the lack of recognition of or even obtuseness to the importance of information and information related managerial actions in the business community. For example, one major study that reviewed a large corpus of work on R&D innovation, [Goldhar et al., 1976], concluded that there are six characteristics of environments that are conducive to technological innovations. The three most important characteristics are all related to the information environment and information flow – specifically: 1) easy access to information by individuals; 2) free flow of information both into and out of the organizations; 3) rewards for sharing, seeking, and using "new" externally developed information sources. Note the 'flow in and out' and the 'sharing, seeking, and using'. Number six is also information environment related, 6) the encouragement of mobility and interpersonal contacts. Yet in a remarkable oversight, the studies' authors never remarked on the dramatic win, place, and show finish of information and knowledge factors.

Another similarly rigorous study [Orpen, C., 1985] examined productivity in R&D intensive electronics/instrumentation organizations. It analyzed various aspects of the behavior of research project managers as perceived by their staff and team members, and it found that in the more productive organizations (as defined by rates of growth and return on assets), the managers were perceived to be significantly more characterized by three aspects of their behavior, all information related: 1) they routed literature and references to scientific and technical staff, 2) they directed their staff to use scientific and technical information (STI) and to purchase STI services, and 3) they encouraged publication of results and supported professional meeting attendance and continuing education. Particularly striking was the finding that not only did information related management behavior tend strongly to discriminate between "high-performance" and "low-performance" companies, but also that *none of the non information related management behaviors measured had any discriminatory value.* Here, given the inability to find any significance for other managerial factors, the failure to remark upon the importance of information and knowledge factors can truly be described as remarkable.

4.4   COMMUNITY-BASED MODELS

The idea of Community of Practice [Wenger and Snyder, 1999], which descends logically from the "Gatekeepers, Information Stars, Boundary Spanners" stream of development has been cited frequently as an important knowledge sharing model. The Community of Practice (CoP) is not necessarily department-based nor centered in one organization. A CoP can consist of those in charge of human resources training, for example, in a number of organizations. These HR professionals can share what they've learned through experience about effective seminar scheduling and working with speakers. Reading a book about effective HR training is one way to learn, but sharing what experienced trainers know is a whole different level of learning. This model is based on the premise that organizational members with similar interests or practices meet to discuss issues of mutual concern and to help each other solve problems. The meeting can often happen in electronic-based forums, and these online discussions are usually self-managing.

The Information Systems literature points to an abundance of KM strategies in the category of Computer Mediated Communication (CMC). Such systems provide the infrastructure for enabling the interactions needed for a group's knowledge synergies and interactive activities [Maier, R., 2002] and may include bulletin boards, electronic meeting/conferencing, or online chat. In this model, the notion of space [Ruhleder, K., 2002], physical or otherwise, is important primarily because the meeting place or system provides an environment that allows for interactions to unfold, at the convenience of individual participants, often asynchronously. Further, such CMC interactions allow for the creation of persistent records [Robins, J., 2002] of the interactions. Chat and other kind of social media transcriptions can be preserved too as another example. To the extent that discourse occurs through such interactions, the dialectics can be archived for future reference and subsequent "reuse." However, as Hislop, D. [2002] points out, while technology may provide the tools for interaction and communication, the application of technology alone may not be a sufficient condition for sustaining the creation and sharing of knowledge.

Another issue with conversational records, Twitter transcripts, and email exchanges as opposed to more traditional knowledge representations in books and journal articles, is that discourse records are not by nature indexed unless a researcher has chosen a set of records for study. A lack of key words, index terms, or metadata on transcriptions and other knowledge aids means that the embedded knowledge can be lost to those who wish to re-use the saved text. Information retrieval and natural language processing have made great advances and may, in the future, be able to help find relevant material, but there are still concerns related to large text, audio and video files that their value may be lost because the content will be forbiddingly difficult or impossible to access.

Group Decision Support Systems (GDSSs) were originally conceived of as collaborative tools where groups came together, participated in brainstorming and then, through human facilitation, voted on items and issues important to the organization. These systems allowed for anonymous voting that moved decisions along rapidly by prioritizing topics more easily than trying to do so without the system's assistance. Participants' knowledge and experience contributed to the democratic process. Another advantage of Group Decision Support Systems, in general, is the ability for each person to speak (through entering opinions via a keypad, or original ideas via a keyboard) anonymously without fear of being politically incorrect or worrying about speaking in opposition to the manager. Contributions could be confidential with the shy on an even plane with the extroverts. A Group Decision Support System (GDSS) is able to calculate the votes and display them graphically, so that an individual attending the meeting can see if she or he were an outlier on certain issues or to determine where his or her vote stood as compared with peers. Although anonymous, each participant can have a unique code, known only to the participant, and follow voting patterns on the graphic display. These systems work well in a face-to-face situation where immediate feedback can be given and displayed. The GDSS has not migrated easily to the Web, however, some web-based systems are available and have adapted to an asynchronous situation. The ability for groups to share knowledge and make decisions using decision technology tools is a beneficial way to combine human know-how and experience with database and display systems. In situations where a range of new products is being considered for development, for example, or a location must be chosen for a new facility as another example, staff can each register opinions and share what they know in order to help make a decision in which they will all be invested because of involvement in the decision making process. These kinds of "invested" decisions cause less friction and are embraced more fully than unilateral ones.

Generic Decision Support Systems (DSS) that act more like expert systems with the added feature of suggesting decision options are well suited to the Web, and they are proliferating as the Web becomes the ubiquitous information and communication platform for information storage and retrieval, and for interaction as well. The range of Web-based DSSs vary in quality from the mundane (e.g., cosmetics or movie choices) to sophisticated tools such as diagnosing illnesses and suggesting appropriate drug therapies. Especially in the medical domain, DSS systems have taken the burden of calculating dangerous drug interactions from the physician's shoulders and offer drug suggestions that include cautions about side effects and problems with other prescriptions patients already have. Anyone can use WebMD or Isabel to test his or her own symptoms against the collected wisdom of clinical experts who have contributed to the systems' "knowledge bases." Generally, these reputable clinical decision support systems are advised by an independent board of medical professionals, and they rely on the experts' knowledge in order to make suggestions for conditions and treatment. Other dependable DSSs have used the expertise of meteorologists to predict storms, knowledge of cattle managers to give advice on culling herds, or the know-how of environmentalists on managing water resources. No doubt, these systems will be replaced by others as technology advances, and their capabilities and functionality will increase.

4.5   REPOSITORY MODEL

The knowledge management repository, a space to store and retrieve knowledge objects has long been a standard in KM programs. It is a model that emphasizes the creation of quality knowledge content in online repositories with re-use as a goal. Markus, M. [2001] argues that the purpose and content of knowledge records in repositories often differ depending on who needs the documentation: the content producer, similar others, or dissimilar others. She emphasizes that a great deal of effort is required to produce quality content, and, as such, part of the burden of documenting and packaging knowledge objects can be transferred to intermediaries, saving time and energy of the organization's staff. In addition, adding context is also another aspect of making content more usable. Markus proposes the roles of human intermediaries in what she terms as "repurposing" of repositories to make them more appropriate for use by others. Examples of activities that could be performed include abstracting, indexing, authoring, and sanitizing or scrubbing content. Because of the costs involved in repackaging and making repository knowledge content more usable to the knowledge seeker, Markus looks to an expanded role for technological support of core competencies of librarians, archivists, data curators, and other information professionals.

Those in the education and training fields may also see great potential in repository and re-use applications. For example, an entire course can be broken down into chunks with the smaller segments presented as just-in-time training on request. Or smaller pieces of a fairly elaborate course can be presented as a workshop or half-day session or even as an abbreviated online webinar with a question and answer session following. Financial firms, IT departments, law firms and others who depend on frequently updated information and new legislative materials are just a few types of organizations that can make good use of the repository and re-use model.

4.6   ACTIVITY-BASED MODELS

While there has been significant work done in terms of Information Systems support for the co-ordination of work [Winograd, T., 1988], the next logical progression would be to link knowledge production and capture with work processes. For example, Blackler, F. [1995] considers knowledge in organizations as socially distributed collective activity systems, and emphasizes the significance of incoherence and dilemma as the key issues in social systems. Similarly, Engeström, Y. [1999] research, using activity systems as cycles of expansive learning in work practices, also points to the importance of activities as providing the necessary context for grounding organizational knowledge.

Based on such a historical-cultural perspective of activity, Hasan, H. [2003] proposed rudiments of a KM system influenced by activity-based models that would link work activities with people and content. Continued development of the model would focus on the motivation of people to contribute content and the meaningfulness of information and knowledge that can be extracted from the contents of such an activity-based system. Incorporating workflow support with a knowledge repository, Kwan and Balasubramanian [2003] take the notion a step further; they propose the design of a KM system they call KnowledgeScope that provides integrated workflow support to capture and retrieve knowledge as an organizational process within the context it is created and used. They also propose a meta-model knowledge structure called Knowledge-In-Context that specifies relationships among processes. The model was implemented with limited workflow functions at a global telecommunications company. While repositories and workflow support have largely developed with limited integration, designs such as this, grounded in case implementations, provide some empirical validity as to the appropriateness and value of incorporating activity as context for knowledge reuse. This emphasis upon context can be seen as part of the maturation of KM as described above in the discussion of stage IV of KM development.

CHAPTER 5

Knowledge "Acts"

5.1 QUESTION ASKING AND ANSWERING

Question asking and answering is a foundational process by which what people know tacitly becomes expressed, and hence, externalized as knowledge. To support such a view, we borrow from speech acts theory [Searle, J., 1969] that amongst others categorizes question asking as a form of a speech act. In adapting the theory, Hirschheim et al. [1995] describe types of speech acts that pertain to aspects of either Knowledge Management (KM), or Information Management (IM). For example, Boahene and Ditsa [2003] suggest that Information Management systems target a base of expressive speech acts by mainly supporting the recall of meaning-attribution while Knowledge Management systems target regulative and constantive speech acts primarily to support the organization and management of dynamic complexity. They reason that IM addresses questions such as 'Where,' 'Who,' 'When,' and 'What,' while KM targets problems involving dynamic complexity, addressing solutions to questions such as 'How' and 'Why.' Quigley and Debons [1999] adopted a similar stance that considers information as texts that primarily answer 'informative' questions such as who, when, what, or where while knowledge is considered as texts that answer more 'explanatory' or 'meaning related' questions such as why or how.

Another category of questions, "What-if," will also fall in the domain of knowledge activity. Since such questions necessitate predicting and prioritizing outcomes, attempts to address such "what-if" questions will require integrating understanding of "what" with "why" and "how" to arrive at reasonable resolution. Thus, what-if questions, primarily seen in the decision making domain, will likely call for exhaustion of all possible scenarios in order to arrive at any "best" alternative. Akin to decision support processes and systems, exercises in "what –if" questions and creating possible scenarios can serve individuals to use existing knowledge and create new knowledge.

5.2 POSTING CONTENT TO REPOSITORIES

Contributing content such as lessons-learned, project experiences, and success stories is another approach to knowledge sharing. The capturing of best practice has often been highlighted as a form of externalized knowledge. O'Dell and Jackson [1998] point out the importance of frameworks for classifying information. For example, they note that Chevron and other groups organize information in their best practice databases using the Process Classification Framework developed by APQC (American Productivity and Quality Council) and Arthur Andersen. Through such a framework, subunits can talk with each other more effectively via a common vocabulary. The authors also point out that the task of entering content into the system should be done by specially appointed people since busy professionals rarely have the time to enter a practice into the database unless it is their job. On the other hand, professionals may not have the time to hand off a document for submission to an appointed surrogate either. For many professionals who are used to online communication and accessing databases and discussion lists, we could argue that it is quicker and easier for the professionals to make the contribution themselves. Nick et al. [2001], noting the importance of learning by experience, point out that experience bases can be developed using case-based reasoning as the underlying concept. However, they also note that experience repositories require continuous maintenance and updating in order to handle continuous streams of experience.

Selvin and Buckingham [2002] describe a tool, Compendium, that claims to support rapid knowledge construction. They ground their claim on an empirical case study of its use in a corporate contingency planning situation by demonstrating the creation of knowledge content in a real time 'on-the-fly' mode of content authoring, complemented by collaborative validation. The authors point out that the product supports both the construction of knowledge as content, or as the collaborative, negotiated, co-constructed approach to verifying and validating content, essentially accommodating both the content and process views of knowledge construction. The developed content is then made available to others for (re)use, or, for re-combination, to support new instances of knowledge creation. Richter et al. [2004] describe a functionally similar tool, TAGGER, designed and operationalized as allowing knowledge acquisition discussions to be "tagged" in real time with the relevant concepts so as to lessen the burden on documentation. As awareness increases for the importance of making knowledge explicit, more and more products will appear to help with creating knowledge bases and decision recommendations, but it is a mindset open to using, sharing, and creating knowledge that will make a difference in creating an organizational knowledge culture.

5.3 (RE)USING KNOWLEDGE

Desouza et al. [2006] assert that the decision to consume knowledge can be framed as a problem of risk evaluation, with perceived complexity and relative advantage being identified as factors relating to intentions to "consume" knowledge. However, it is essential that the knowledge consumer is able to reasonably frame his or her knowledge needs. Belkin et al. [1982] found that during problem articulation, users have anomalous states of knowledge, and they may not be able to specify their information needs accurately. Since the publication of this seminal work legions of researchers have worked on systems that will help people formulate effective questions that will retrieve relevant information. McMahon et al. [2004], studying team work involving engineering design, suggest that both codification and personalization approaches to knowledge reuse are relevant. They recognize the notion of information value, allowing for the matching of information to the knowledge needs of the user. They propose that good representations of both information characteristics and user characteristics are essential.

For such a purpose, Hicks et al. [2002] propose a practical framework for the requirements of capturing, storing, and reusing information and knowledge in engineering design. They distinguish between the process that generates knowledge and the knowledge element generated by the process.

They argue that the form in which knowledge is captured has to be informed by the eventual application, or reuse of the content. That is, echoing McMahon et al. [2004], the needs of the user should drive the knowledge capture process.

5.4   KNOWLEDGE-BASED DECISION MAKING

In general, decision making involves identifying alternatives, projecting probabilities and outcomes of alternatives, and evaluating outcomes according to known preferences and implications for stakeholders. Choo, C. [2002] suggests that decision making activity requires the establishment of shared meanings and the assumption of prior knowledge.

Shared meanings and purposes as well as new knowledge and capabilities, converge on decision making as the activity leading to the selection and initiation of action. Shared meanings, agendas, and identities select the premises, rules, and routines that structure decision making. New knowledge and capabilities make possible new alternatives and outcomes, expanding the range of available organizational responses [Choo, C., 2002, p. 86]. Choo further proposes that information flows are a central process that bridges knowledge creation and decision making activity. Information flows continuously between sense making, knowledge creating, and decision making, so that the outcome of information use in one mode provides the elaborated context and the expanded resources for information use in the other modes [Choo, C., 2002, p. 85].

Information used in one activity that results in new knowledge will, in turn, be used to guide selection of alternatives in future tasks that involve decision making. Codified rules and routines would be relied on to support evaluation of alternatives and selection of action decisions. Choice of alternatives, and decision outcomes then provide the backdrop upon which sense making, or justification, of decision rationale occurs. Such decision rationale, and its associated sense making can then be codified for (re)use in other contexts, applied to future activities that draw on it to create new instances of knowledge.

In such decision oriented activity, we have proposed that "what-if" questions are the dominant type of speech act performed. Support for such scenario predicting questions will demand rich context upon which to apply knowledge of the past and the present to bear on the problem or situation at hand. We would like to refocus the discussion of knowledge management strategy to the demands of complex, dynamic, contextual, and emergent decision processes. The next section of the paper gives an overview of knowledge management processes, procedures, and practices. Chapter 8 discusses how they are related to decision making.

CHAPTER 6

Knowledge Management in Practice

Knowledge management or knowledge sharing manifest themselves in many ways in the workplace; that may include ordinary events, such as facilitated meetings or informal conversations or more complex interactions that require information and communication technology. Since building knowledge may require the analysis and synthesis of information, the lines between working with information and working with knowledge or knowledge artifacts easily become blurred. Under the aegis of 'knowledge management', there are three types of processes that are generally considered to be essential: finding or uncovering knowledge [Ehrlich, K., 2003, Learn, L., 2002, Zack, M., 1999], sharing knowledge [Ackerman et al., 2003], and the development of new knowledge [Argyris and Schon, 1978, 1996, Baumard, P., 1999, Harvard Business Review, 1998]. All may play a role in assisting with decision making and encouraging innovation. The next section of this paper will focus on these categories of practice and will give examples and references to each.

6.1   KM IN PRACTICE - PROCESSES

A very useful way of thinking is to conceptualize KM as the actualization of what Powell, T. [2001a] calls the "Knowledge Value Chain." The chain is straightforward, a pyramid, in fact, leading from Data at the bottom through Information, Knowledge, Intelligence, Decision, and Action, to Value. The notion is simple, but the explication is sophisticated and complex. Value to the organization is ultimately what KM is about.

6.1.1   FINDING INFORMATION AND KNOWLEDGE

Finding information and knowledge refers to processes that allow organizations to make sense and make use of data, information, and knowledge objects that may be present but are not codified, analyzed, nor accessible to members. Knowledge exists in all organizations, but all knowledge may not be explicit. Knowledge objects or artifacts are entities that represent knowledge existing within organizational members [McInerney, C., 2002]. A long-time employee may have a deep understanding of processes and guidelines, but he or she may never have written them down or compiled them in a document like a procedural manual. As a gatekeeper of know-how about "how things are done," this worker has the power of knowing, but may also find that it is tedious and annoying having to answer the same questions over and over again. Codifying his or her knowledge in a manual would allow others to read and understand how tasks and projects can proceed to completion.

Another example of finding knowledge is a situation where vendor information is available, but no one in a firm has evaluated vendors nor kept a record of the evaluations and lessons learned about working with vendors, nor has made explicit some of the vagaries of a specific vendor's organizational and decision making structure. The person who deals with vendors may have extensive knowledge about them, but when he or she leaves the firm, the knowledge goes away [Davenport and Prusak, 1998a]. When a staff member intentionally and systematically compiles a database of vendors with accompanying details in a relational database, reports can be produced that allow managers to see a pattern of purchasing, turn-around time, and staff ratings of vendors. Trends can be examined and analyzed so that new understandings of procurement practices and purchasing can be made explicit. The knowledge that once existed only within one person can be used (at least to some degree) by others who find it represented, codified, and organized in electronic form.

One aspect of finding and dissemination of information is the organization of knowledge objects so that they can be found easily. Assigning index terms, tagging or, in the case of an intranet and the Web, metadata, allows ease of retrieval [Learn, L., 2002]. Digitally recorded presentations, brochures, reports of lessons learned, and best practices can all be made accessible through careful indexing and intelligent information architecture [Morville, P., 2005, Rosenfeld and Morville, 2002].

6.1.2   SHARING INFORMATION AND KNOWLEDGE

Sharing of information for knowledge development is the most traditional collection of processes, easily understood, but often overlooked in a systematic knowledge management program. Sharing refers to the willingness and ability of the knowledgeable to share what they know to help others expand their own learning and knowing. Teaching and learning activities, such as online universities in industry, mentoring programs, apprenticeships, and training programs all serve as opportunities for individuals to share knowledge. The live interactions that occur in lectures and other kinds of learning sessions can now be captured fairly easily with digital video or audio equipment. Even mobile devices have these capabilities. They can then be indexed and placed on a shared file platform or in an intranet. If indexed appropriately, knowledge workers can find the audio and video and use these things over and over again. Especially for mobile workers who need to travel, be onsite for client meetings, or work from home, having access to learning opportunities through multi-media files can help them stay in tune with the organization and on-site workers. Chait, L. [2008] makes a very important point about the readiness of people to share knowledge. Based on his experience at Arthur D. Little, he found that the principal reason for reluctance on the part of key players to put their knowledge into a lessons learned database was a concern that the lesson learned might be misapplied if the congruence, or the lack of, between the context of the area from which the lesson was derived and the context of the intended application area was not well understood. The principals therefore wanted the person who needed the information or knowledge to have to come to them, so that the two contexts could be discussed and the applicability properly understood. The principles were, in general, quite willing to have it be broadcast that they had a lesson learned in a particular area, but in many cases, they did not want so much to be revealed that someone else would feel that they knew enough about that lesson to take it and run with it without consultation first.

6.1.3   DEVELOPMENT OF KNOWLEDGE

Knowledge development takes place when individuals work to create new understandings, innovations, and a synthesis of what is known already together with newly acquired information or knowledge. Although individuals can intentionally develop their own knowledge through seeking opportunities to be creative and learn, the development of knowledge is often a social process. Meetings, teleconferences, planning sessions, knowledge cafes, and team think tank sessions all serve to help workers develop knowledge together. The synergies brought about by effective meetings can encourage the development of new knowledge. Allowing individuals to take risks and occasionally make mistakes (and learn from them) can also develop a culture of innovation that fosters the creation of new knowledge through research and experimentation. The rise of the information economy in the United States in the late twentieth century is often attributed to the ongoing incentives and funding given to middle class and poor families in the decades after World War II [Killing off the American future, 2006]. Similarly, other countries such as Ireland rose to prominence and gained enormous economic ground around the time of the millennium due to long standing policies and practices of making high quality and affordable education available to all [Friedman, T., 2005].

6.2   KM IN PRACTICE - PROCEDURES AND PRACTICES

Note that KM is a complex topic, and in attempting to write about its various dimensions and to address it from different perspectives, some overlap is unavoidable.

6.2.1   KNOWLEDGE AUDIT

The obvious first step in launching a formal KM program throughout an organization is to conduct an information or knowledge audit. An audit answers the questions of what information and knowledge exists in the organization and where is it? Who maintains it? Who has access to it? Etc.

The idea of an information auditory much predates KM as we have defined KM here. Accompanying, or more accurately a component of, the Information Resources Management (IRM) movement of the 1970's was a strong emphasis upon the information or knowledge audit. The foremost exponent of the information or knowledge audit was Forrest (Woody) Horton. He and Burk developed a program called 'InfoMapper' [Burk and Horton, 1988] precisely to facilitate the conduct of an information audit. Auditing as it was then defined focused primarily on what data was formally captured in documents and databases. At that time, the Internet and Web portals did not yet exist, and there was a very legitimate concern that data was being captured in an unplanned and decentralized fashion and that the data was held as the "slave of the program." Furthermore, the fear was that the data was not being well documented that it was being was stored in silos and that valuable data and information, whose very existence was sometimes known only to a few people was often unknown to many of those who could benefit from using it. While tacit or implicit information was not ignored, the emphasis was very much upon explicit captured data and information.

Some of the reasons for and benefits of an information audit include:

First of course, the elucidation of what information the organization possesses: where it is located? how is it organized? how can it be accessed? who is responsible for it? etc.

In addition:

The identification of duplicate or partially duplicated information and information gathering and maintenance, with the potential realization of cost savings.

The identification of information being gathered and maintained that is no longer salient or necessary, with the potential realization of cost savings.

Note that the two last points above are often more persuasive in selling the need, or opportunity, for an information audit, to management than is the argument for greater and better access, which in the long run is really the most compelling reason.

The identification of information lacunae or gaps. Asking 'What if?' and 'What about?' questions is apt to bring gaps to light. Then questions can be asked and answered concerning the availability, cost, and potential utility of obtaining that information.

With the development of KM, there ensued a shift to a much greater emphasis upon knowledge embodied in people. Indeed, Moulton, L. [2008] advocates a three-stage process for a knowledge audit that starts with people and emphasizes knowledge embodied in people. The first stage focuses on people, "their knowledge and expertise and their connections to others" [Moulton, L., 2008, p. 80]. The ideal result is a "map" of:

Who is connected to whom, formally and informally?

What are their formal roles and job descriptions, and informal relationships and roles?

Where do expertise, methods, differing views of the organization reside?

What are the successful knowledge sharing engagements and practices?

What are the barriers to information and knowledge transfer?

What are the cultural behaviors that are dictating successes or failures to share and leverage knowledge?

Clearly, the techniques used in creating a knowledge audit or knowledge map are those borrowed from social network analysis and anthropology, and appropriately so, since Knowledge Management is interdisciplinary by nature, spanning boundaries of thought and interests.

The second stage focuses on programs, projects, and products. How does information flow in and around them? Are there communities of practice even if interest groups are not so named? The third stage focuses on documentation and how information is captured explicitly. The KM era notion of an information audit, in contrast with the earlier IRM era, is definitely focused on people first. In fact, Moulton's third stage of the knowledge audit is essentially the traditional information audit, with a Stage one and a Stage two added in front. Powell, T. [2004b] provides a "Knowledge Matrix" that serves as a very useful checklist for a knowledge audit.

However, all of the points behind a "classic" information audit remain just as important as ever. What the literature and the awareness concerning Intellectual Capital and KM have done is to give a new importance to information and knowledge and to create a management climate that is apt to be much more receptive to the idea of an information and knowledge audit. It's critical for all involved in such an endeavor to remember that knowledge grows from information, so careful oversight of information is necessary as a foundation for knowledge development and the formation of a knowledge sharing culture.

6.2.2   TAGS, TAXONOMIES, AND CONTENT MANAGEMENT

Having identified and located information and knowledge, the obvious next step is to make it relocatable and retrievable, made possible by tagging and creating taxonomies. (Note that the term used by far the most frequently in this context in KM is "taxonomy." The traditional professional information community would call what most authors in the KM field call a taxonomy a classification scheme, or a classificatory or syndetic structure. But most writers in the KM domain come from the business world and are unaware of that terminology, and use the word "taxonomy" that they remember from their high school and college science courses.) Stage III of the development of KM, described above, can well be called the Taxonomy Stage.

The tag and taxonomy stage of KM consists primarily of assembling various information resources in some sort of portal-like environment and making them available to the organization. This can include internally generated information, including lessons learned databases and expertise locators, as well as external information, the open web and also deep web information subscribed to by the organization. With the arrival of extensive email use by virtually all organizations the extent of internal information to be managed has exploded. The Enron scandal and consequent Sarbanes-Oxley legislation in the US had made the retention and management of electronic information mandatory, not optional. Increased use of social media within the organization has expanded the domain of information to be managed still further. This massive increase in information interaction, including use of digitized video and audio and the organization's own web pages has resulted in the development of what is a major subfield within KM, that of "Content Management" or "Enterprise Content Management." The area is also frequently labeled as CMS, Content Management Systems. There are numerous vendors competing to sell content management systems, or "Enterprise Search" systems as they are also termed. Emerging fields or technologies frequently have overlapping and competing terminology before the field matures and a consistent terminology emerges. The area of managing content is still in its early days and will clearly expand and develop as organizations see the need for preserving, organizing, and re-using knowledge objects.

The obvious consequence of this plethora of data and information from multiple sources is great terminological inconsistency and confusion, and that, in turn, drives the appeal of syndetic data structures and taxonomies that can assist the user in locating information or knowledge and result in better and more effective searching. There is now an industry sector whose role is to provide software and expertise to assist organizations in developing their taxonomic systems. Most of those organizations are represented among the vendors at the KM World Conference. KMWorld, a controlled circulation, i.e., free, magazine publishes a very useful annual compilation of vendors and products, particularly in the CMS domain, but including KM broadly as well.

An interesting token of the increased emphasis paid to this subject is that since 2006 the KM-World Conference has been accompanied by a two-day well received and well attended "Taxonomy Boot Camp."

6.2.3 LESSONS LEARNED DATABASES

Lessons Learned databases are databases that attempt to capture and to make accessible knowledge that has been operationally obtained and typically would not have been captured in a fixed medium (to use copyright terminology). In the KM context, the emphasis is typically upon capturing knowledge embedded in persons and making it explicit. The lessons learned concept or practice is one that might be described as having been birthed by KM, as there is very little in the way of a direct antecedent. Early in the KM movement, the phrase typically used was "best practices," but that phrase was soon replaced with "lessons learned." The reasons were that "lessons learned" was broader and more inclusive, and because "best practice" seemed too restrictive and could be interpreted as meaning there was only one best practice in a situation. What might be a best practice in North American culture, might well not be a best practice in another culture. The major international consulting firms were very aware of this and led the movement to substitute the new term. "Best Practices" succeeded by "Lessons Learned" was the most common hallmark phrase of Stage I of KM development.

One possible antecedent for "Lessons Learned" was the Hepatitis Knowledge Base [Bernstein et al., 1980] developed in the 1970's by the National Library of Medicine. The Hepatitis Knowledge Base, however, was based on a vetting of the formal journal literature; it was not based on tacit or implicit knowledge. Another possible antecedent is the World War II debriefing of pilots after a mission. The primary purpose was to gather military intelligence, but a clear secondary purpose was to identify lessons learned, though they were not so named, to pass on to other pilots and instructors. Similarly, the U. S. Navy Submarine Service after an embarrassingly lengthy fiasco of torpedoes that failed to detonate properly, and an even more embarrassing failure to follow up on captains' consistent torpedo failure reports, instituted a system of widely disseminated "Captain's Patrol Reports" with the intent of avoiding any such fiasco in the future. The Captain's Patrol Reports were very clearly designed to encourage analytical reporting, not just reports of actions taken and ships sunk, but reports with reasoned analyses of the reasons for failure and success, and emphasizing that a key purpose of the report was to make recommendations about strategy for senior officers to mull over, and about tactics for other skippers to take advantage of.

The military has become an avid proponent of the lessons learned concept. The phrase the military uses is "After Action Reports." The concept is very simple: don't rely on someone to make a report. There will almost always be too many things immediately demanding that person's attention after an action. It should be the responsibility of one person to create the report, separating the wheat from the chaff, and then the lessons learned can be disseminated. The concept is by no means limited to the military. Prusak, L. [2004] states that the greatest KM failure and missed opportunity is that so often in the corporate world the project team is disbanded, and the team members reassigned before there is any debriefing or after action report assembled. Organizations operating in a project team milieu need to pay very close attention to this issue, and to set up an after action procedure with clearly delineated responsibility for its implementation.

A wonderfully instructive example of a "lesson learned" is recounted by the KM consultant Mazzie, M. [2003]. The story derives from his experience in the KM department at Wyeth Pharmaceuticals. Wyeth had recently introduced a new pharmaceutical agent primarily for pediatric use. They expected it to be a substantial success because, unlike its competitors, it needed to be administered only once a day, which would make it much easier for the caregiver to ensure that the child followed the drug regimen. Sales of the drug started well but soon turned disappointing. One sales rep (what the pharmaceutical industry used to call detail men), however, discovered the reason for the disappointing sales and the solution. The problem was that kids objected strenuously to the taste of the drug, and caregivers were reporting to prescribing physicians that they couldn't get their kid to continue taking the drug. The solution was orange juice. A swig of orange juice quite effectively masked the offensive taste. If the sales rep informed the physician that the therapy should be conveyed to the caregiver as the pill and a glass of orange juice taken simultaneously first thing in the morning, then there was no dissatisfaction and sales were fine. There are also lessons learned in this story about motivation for information sharing (discussed later).

The implementation of a lessons learned system is complex both politically and operationally. Many of the questions surrounding such a system are difficult to answer. Who is to decide what constitutes a worthwhile lesson learned? Are employees free to submit to the system unvetted? Most successful lessons learned implementations have concluded that such a system needs to be monitored and that there needs to be a vetting and approval mechanism before items are mounted as lessons learned. How long do items stay in the system? Who decides when an item is no longer salient and timely? Most successful lessons learned systems have an active weeding or stratification process. Without a clearly designed process for weeding, the proportion of new and crisp items inevitably declines, the system begins to look stale, and usage and utility falls. It is the same phenomenon that school librarians have observed for decades. Materials need to be current and relevant. Deletion, of course, is not necessarily loss and destruction. Using stratification principles, items removed from the foreground can be archived and moved to the background, but still be available.

These questions need to be carefully thought out and resolved, and the mechanisms designed and put in place before a lessons learned system is launched. Inattention can easily lead to failure and the tarring of subsequent efforts.

6.2.4   EXPERTISE LOCATION

If knowledge resides in people, then one of the best ways to learn what an expert knows is to talk with one. Locating the right expert with the knowledge you need, though, can be a problem. The basic function of an expertise locator system is straightforward, it is to identify and locate those persons within an organization who have expertise in a particular area. Such systems were commonly known as "Yellow Page" systems in the early days of KM, the name coming from the telephone book yellow pages, the section of the phone book, or a separate volume of the phone book, organized for subject search. In recent years, the term expertise locator or expertise location has replaced yellow pages as being rather more precise. After all the yellow pages metaphor with its implication of subject search could apply to many areas of KM, such as for example lessons learned and content management.

Expertise location systems are another aspect of KM that certainly predates KM thinking. The Mitre Corporation, for example, developed such a system in 1978. It was based upon creating a database developed from reformatted resumes retrieved from word-processing tapes, and upon the development of a competence area thesaurus to improve retrieval. However, even in a technologically sophisticated organization, generously, by the standards of the time, supplied with computer work-stations, the system was underutilized, fell into disuse, and was for all practical purposes abandoned. More than two decades later, the system was essentially reinvented as part of a larger KM initiative [Mattox et al., 1999], by a development team that was totally unaware of the previous system and its thesaurus, which they would have called a "taxonomy," and which had to be recreated. A journal article on the history of KM at Mitre [Maybury, M., 2003] starts its discussion in the late 1980s and makes no mention of the 1978 system. This is a good example of valuable organizational knowledge and expertise being lost due primarily to normal personnel turnover.

There are now three areas which typically supply data for an expertise locator system, employee resumes, employee self identification of areas of expertise, typically by being requested to fill out a form online, or by algorithmic analysis of electronic communications from and to the employee. The latter approach is typically based on email traffic, but it can include other social networking electronic communications such as Twitter and Facebook. Commercial packages to match queries with expertise are available. Most of them have load-balancing schemes so as not to overload any particular expert. Typically, such systems rank the degree of presumed expertise and will shift a query down the expertise ranking when the higher choices appear to be becoming overloaded. Such systems also often have a feature by which the requester can flag the request as a priority, and the system will then try to match higher priority requests with higher presumed (calculated) expertise rank.

6.2.5 COMMUNITIES OF PRACTICE (COPS)

Communities of Practice (CoPs) are groups of individuals with shared interests that come together in person or virtually to tell stories, discuss best practices, and talk over lessons learned [Wenger, E., 1998a, Wenger and Snyder, 1999]. Communities of practice emphasize the social nature of learning within or across organizations. Conversations around the water cooler are often taken for granted, but organizations find that when workers give up a company office to work out of their home, that the natural knowledge sharing that occurs in social spaces must be replicated in an online form. As an alternative, workers are called in for periodic meetings for the express purpose of learning from each other [McInerney, C., 2000], or they're encouraged to participate on online forums. In an information society where knowledge is considered an important resource for individuals and organizations, processes to share knowledge should be considered integral to any strategic or tactical plan. Many large firms, especially those in the pharmaceutical industry, have adopted internship and leadership programs for the express purpose of helping promising new associates learn about how the company operates and to teach them that today, learning is an ongoing part of work.

In the context of KM, CoPs are generally understood to mean electronically linked communities. Electronic linkage is not essential of course, but since KM arose in the consulting community from the awareness of the potential of Intranets to link geographically dispersed organizations, this orientation is understandable and inevitable.

The organization and maintenance of CoPs is not a simple and easy undertaking. As Durham, M. [2004] points out, there are several key roles to be filled, which she describes as manager, moderator, and thought leader. They need not necessarily be three separate people, but in some cases they will need to be.

For a CoP, some questions that need to be thought about are:

Who fills those roles? manager, moderator, and thought leader.

How is the CoP managed?

Are postings open, or does someone vet or edit the postings?

How is the CoP kept fresh and vital?

When and how (under what rules) are items removed?

How are those items archived? (Stratification again)

Who reviews the CoP for activity? Identifies potential for new members, or suggests that the CoP may have outlived its usefulness?

6.3 PROCESSES, PROCEDURES, AND PRACTICES MATRIX

If we create a matrix in which the rows are KM Processes and the columns KM Procedures and Practices, and in which the ordering, top to bottom and left to right is roughly in chronological or developmental order, and we check which process a practice or procedure is *primarily* designed to serve, the matrix looks like:

	K Audit	Tags, T, & CM	Lessons L.	Expertise. Loc.	CoPs
Finding	X	X X X			X
Sharing			X X		X
Developing				X X	

Figure 6.1: Processes and Procedures & Practices Matrix.

That matrix reveals several interesting things. Almost everything one does in KM is designed to help find information and knowledge. However, if we assume that the main goal of KM is to share knowledge and even more importantly to develop new knowledge, then the Knowledge Audit and the Tags, Taxonomies and Content Management stages are the underpinnings and the tools. It is the knowledge sharing and knowledge creation of one on one communications enabled by expertise locators, and the communal sharing and creation of knowledge enabled by communities of practice toward which KM development should be aimed.

CHAPTER 7

Knowledge Management Issues

7.1 EXPLICIT & TACIT; BUT BETTER: EXPLICIT, IMPLICIT, & TACIT

In the KM landscape, Knowledge is most commonly categorized as either explicit or tacit. This characterization is rather too simple, but a more important point, and a criticism, is that it is misleading. A much more nuanced and useful characterization is to describe knowledge as explicit, implicit, and tacit.

Explicit: information or knowledge that is set out in tangible form.

Implicit: information or knowledge that is not set out in tangible form, but could be.

Tacit: information or knowledge that one would have extreme difficulty operationally setting out in tangible form.

The classic example in the KM literature of tacit knowledge is Nonaka and Takeuchi's example of the kinesthetic knowledge that was necessary to design and engineer a home bread maker, a knowledge that could only be gained or transferred by having engineers work alongside bread makers and learn the motions and the 'feel' necessary to knead bread dough [Nonaka and Takeuchi, 1995]. Prusak, L. [2001] uses the example of reading Ted Williams' "The Art of Hitting" in an attempt to improve his little league batting skills, and of how little of Ted Williams' knowledge could, in fact, be made explicit or transferable.

Implicit knowledge is that which has not been made explicit, but could be, an example might be: "One might think that from organization X's structure and their organizational chart that for this area Smith would be the decision maker, but in fact it is Menendez."

The danger of the explicit-tacit dichotomy is that by forcing knowledge into two categories, i.e., explicit, that which is set out in tangible form, and tacit, that which is in within people, it is easy to think in terms of explicit knowledge which calls for 'collecting' KM methodologies, and tacit knowledge, which calls for 'connecting' KM methodologies, and to overlook the fact that in many cases, what may be needed is to convert implicit tacit knowledge to explicit knowledge, and that both collecting and connecting KM methodologies may be appropriate.

One of the most compelling presentation of this point is made by Keen and Tan (2007), from whom the terminology of explicit, implicit, and tacit is adopted, but others, particularly Hawamdeh and Refai [2008] and Wilson, T. [2002], make the same point. Hawamdeh uses the term "articulated knowledge" for what is referred to above as "implicit." Wilson points out that the "guilty party" for the false premise that truly tacit knowledge could be captured in tangible form was Nonaka, that despite his excellent example of truly tacit knowledge, he conflated tacit and "expressible knowledge," Wilson's term for what we have called "implicit."

7.2   TRAINING AND USER EDUCATION IN KM

It is now common to observe, and indeed it is a theme of this book, that although KM and the recognition of its importance was precipitated by the appearance of the internet and its brethren, intranets and extranets, that fundamentally KM is more about people and organizational culture than it is about technology.

An important aspect of that human resource side of KM that is far too often not adequately recognized is the need for user support and user education and training to make KM work effectively and to be successful. A well known KPMG Consulting [2000] study, one of the largest studies ever done on the implementation of KM systems, covering more than 400 firms, provides fascinating and compelling documentation in regard to this problem. They reported that of the 288 firms that had KM in systems in place or were setting up such a system, there were 137 cases, nearly half, where the benefits failed to meet expectations. The breakdown of why, from the company's perspective, the benefits failed to meet expectations was as shown in Figure 7.1.

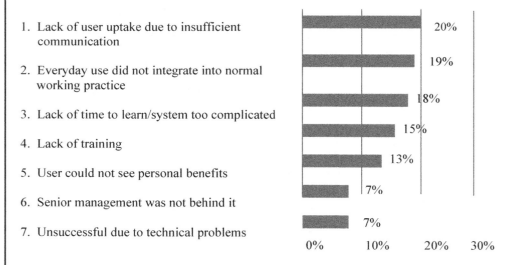

1. Lack of user uptake due to insufficient communication — 20%

2. Everyday use did not integrate into normal working practice — 19%

3. Lack of time to learn/system too complicated — 18%

4. Lack of training — 15%

5. User could not see personal benefits — 13%

6. Senior management was not behind it — 7%

7. Unsuccessful due to technical problems — 7%

Base: all where benefits fail to meet expectations (137)

* Note these percentages add to 99% due to rounding error; there is no overlap

Figure 7.1: Reasons for KM failure as identified by KPMG.

What is striking about these data, and striking on two levels, is that:

#1) lack of user uptake due to insufficient communication,

#2) lack of time to learn/ system too complicated, and

#3) lack of training,

are all fundamentally the same reason - inadequate training and user education. With that recognized, the table can be recast in a much more informative fashion:

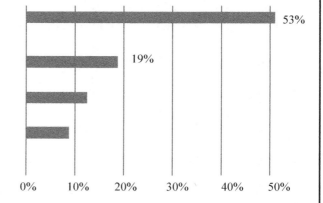

1. Inadequate training and user education

2. Everyday use did not integrate into working practice

3. User could not see personal benefits

4. Senior management was not behind it

5. Unsuccessful due to technical problems

Figure 7.2: Reasons for KM failure as Recast.

It is also quite likely that some component of reason 5, "Unsuccessful due to technical problems," is also reflective of inadequate training and user education, so that the 53% may in fact be an underestimate.

What is notable at the first level is that inadequate training and user education is by far the most prominent reason for why benefits failed to meet expectations, accounting for the majority of failures, exceeding all other reasons combined.

One small caveat is that as Davenport and Prusak [1998a, p. 26] observe, sometimes lack of training may take the blame when the real culprit is a combination of naïve expectation and a failure to adequately address the corporate cultural changes needed. Even with this factored in, however, the predominance of the training and user education factor is striking.

Second, and very surprising is that the KPMG report fundamentally fails to pick up on this rather dramatic finding. To their credit, KPMG does observe that "These responses confirm the fundamental flaw in viewing KM as a technology issue: it is not the technology that is holding organizations back but a lack of strategy and a failure to build KM in the organization's day-to-day operations and its culture in order to encourage end-user buy-in," but that is as far as they go toward a recognition of what is the most central and dramatic finding of their study, the importance of user training and education.

7.2.1   USER TRAINING AND EDUCATION VERSUS MANAGEMENT SUPPORT FOR KM

An interesting comparison is that in KPMG's data, the percent that reports that inadequate training and user education was the principal problem for failure was 53%, while the percent that reports that "senior management was not behind it" was the principal problem for failure was only 7%.

It is not the case that senior management support is unimportant, but consider and contrast the proportion of the KM literature that emphasizes the key importance of getting senior management support with the proportion of the KM literature that emphasizes the key importance of setting up adequate and extensive support for user education and training, and one immediately gets a feel for the extent of the Achilles' heel problem. The culture of KM has clearly not been adequately aware of the importance of training and user education, and the KPMG report illustrates and illuminates the problem in a wonderfully compelling fashion by, ironically enough, not recognizing it. One could argue that a system should be intuitive enough not to require much education or training to use it effectively, but that may be easier to say from today's perspective when the development and use of Web-based technology platforms are more common than they were during the time when KPMG conducted its study.

7.3   TRUST IN KM

A concern that occurs commonly in KM circles is that of trust, in other people and trust in and for the organization. Prusak, L. [1998] has long bewailed the fact that companies downsize, then realize that a great deal of knowledge and competitive advantage has departed with those downsized. The same companies may then hear about KM and decide that KM will be the panacea for the problems created by departed knowledge, deciding to implement a KM program at precisely the wrong time, when trust in the organization is at its lowest ebb.

Trust between individuals has been rather thoroughly researched. Ridings et al. [2002] describe a model of trust that they believed occurred in virtual community networks (CoPs). In this model, there are three inputs that lead to trust and to the outputs of the desire to get information and the desire to give information. The three inputs are (1) perceived responsiveness, (2) others confiding personal information, and (3) disposition to trust. Perceived responsiveness is the notion that individuals will eagerly give more information and desire to get information if others respond to them in the network. The second input is about relieving the feelings of not knowing the people that are in the network. Since the faces are not seen, there may be a feeling of talking with strangers (or they may actually be strangers). By people giving personal information such as age, gender, or something personal about themselves, they are making themselves less of a "stranger" to the network participants.

The third input is the disposition to trust. More specifically it has been defined as the, "extent that she or he demonstrates a consistent tendency to be willing to depend on others across a broad spectrum or situations and persons" [McKnight et al., 1998, p. 477]. Some people are just more trusting than others, and this may come from experiences with the world around them including their interactions with their peers, parents, and authority figures [Rotter, J., 1967]. Positive experiences build our disposition to trust while negative experiences break our disposition to trust [Rotter, J., 1971]. Many researchers have confirmed that more knowledge will be transferred when there is more trust between the users of the virtual community [Chiu et al., 2006, Hsu et al., 2007, Inkpen and Tsang, 2005, Lin et al., 2009, Yu et al., 2009]. However, Renzi found that the reason trust improves knowledge transfer is due to the reduction in fear [Renzl, B., 2008]. This is what Zhang and colleagues called 'psychological safety' [Zhang et al., 2010].

KM authors repeatedly stress the importance of trust in the KM environment, particularly in the context of CoPs. Malhotra, Y. [2004], for example, in one of the better such articles, discusses the importance of establishing trust and commitment in implementing KM. Davenport, E. [2004] also emphasizes the importance of such trust, and intriguingly reports that the OPAL (Online PArtner Lens, an EC funded project) team was at that time considering the possibility of establishing an HTML flavor, "Trust ML" (Markup Language) that would formalize attributes of trust made manifest in electronic interactions. No subsequent developments ensued, however.

KM failure is often attributed to lack of trust. Davenport and Rasmussen (2008) give an excellent such example of failure in trust at a Scottish public service agency. The reason for the difficulty in doing such trust research is that very few organizations like to be studied in this regard. It is hard, for example, to imagine Pfizer Inc. or other pharmaceutical firms welcoming a study as to why employee trust in the organization has so precipitously declined in the last two decades.

And, while it is standard advice that a KM effort should be spearheaded by a senior executive with line experience along with confidence and trust of senior management in the project, it is still the case that the KM undertaking will have very little opportunity or leverage to change the level of trust in the organization in the short term. The KM team will have to accept the level of trust in the organization as a given, at least for some time. The fact that increased communication, knowledge sharing and greater openness may ultimately lead to greater trust in the organization does constitute a justification for implementing KM.

7.4   INFORMATION SECURITY VERSUS INFORMATION SHARING

KM is about knowledge sharing, but organizations have legitimate concerns about protecting the confidentiality and security of key proprietary information, and these two interests – security and sharing – create a tension. For not-for-profit quasi-governmental organizations such as the World Bank, the tension may not be as strong, and they can easily define as part of their mission to create and share information. Indeed, James D. Wolfensohn, the president of the World Bank from 1995 to 2005, did precisely that, by stating clearly that the principle product of the World Bank was not loans, but rather it was the information and knowledge which was created by loaning money for and facilitating development projects. The knowledge learned would facilitate and could be applied for further developments, and that was the real product and the goal of the World Bank's efforts [Srikantaiah, T., 2004]. Not surprisingly, the World Bank was a pioneer in applying and publicizing KM. In particular, the World Bank promoted and publicized the utility of communities of practice (CoPs) for knowledge sharing, establishing communities of practice that included not only bank employees, but also others from around the world. If the subject of the CoP was road building in arid conditions, then it made a conscious effort to recruit CoP members from, for example, the Australian Road Research Board and the Arizona Department of Transportation.

For-profit organizations in a competitive environment unfortunately cannot be so sanguine about knowledge sharing, particularly if their product, such as pharmaceuticals, depends upon patent protection and being the first to patent. However, as pointed out in the section above on "KM as the Extension of the Successful R&D Environment," the most successful pharmaceutical companies in terms of drug development are the companies whose researchers characterize themselves as the least concerned in the industry about protecting the confidentiality of the company's proprietary information. Very clearly, the more successful a company's drug development efforts were, the more it was characterized by a style of rich, deep, and open communication, both within the organization, but also with the external environment.

The lessons learned here for KM seem clearly to be that the balance between protection and security versus promoting openness needs to be very clearly considered, and that if it's necessary to err, err on the side of rich, deep, and open communication.

7.5   KM AND COMPETITIVE INTELLIGENCE

There is an obvious synergy between KM and Competitive Intelligence. Competitive Intelligence (CI) is of no use unless it is deployed and used. That deployment and use is, of course, or at least should be, a component of KM. Much, indeed most of the information or knowledge gathered in competitive intelligence, comes from the open literature [Combs and Moorhead, 1993]. This fits neatly with what we have described as Stage IV of KM development, the awareness of the importance of information and knowledge external to the organization. Furthermore, the intelligence that does not come from the open literature, which is what comes to mind to most people when they hear the phrase CI, is in fact the same sort of information gathering, a flavor of after action reports, if you will, as exists in KM. (The CI community is always at great pains to point out that CI is not espionage.) While KM and CI have to date been rather separate communities, there has begun to be increasing convergence [Maag and Flint, 2004, Shelfer, K., 2004]. Benchmarking and Gap Analysis are two areas in particular where Shelfer, K. [2004] points out that KM and CI engage in highly overlapping activities.

The issue of KM and CI also relates very closely to the Issue of "Information Security vs. Information Sharing." Barth, S. [2004] points out, in the context of planning for KM, that CI is not just about information gathering (offense), it is also about protecting one's organization from the competitions CI (defense). Therefore, particularly in for-profit organizations, both KM and CI have a common problem area, how to balance "rich, deep, and open," with information and knowledge security. Clearly, the practices and policies that result from addressing this issue need to be coordinated, and clearly therefore KM and CI need to be coordinated.

The obvious bottom line is that KM and CI should not be "silos." The more integrated they are the better, and if they are to be integrated, then since KM is a much broader concept than CI, the most straightforward relationship is for CI to be part of KM. But, since in many organizations the concept of CI, and a structure to support it, has been around rather longer than the KM concept, this may require some delicate footwork to implement. In many cases, the best approach is to point out the overlap, negotiate the necessary coordinating mechanism, and then let things evolve toward a fuller integration.

7.6   KM MOTIVATION (CREATING MOTIVATION FOR KM)

The issue of how to motivate people to take part in the KM process and to share knowledge is an important one.

One obvious answer is to use KM participation and knowledge sharing as one measure of performance review and, hence, related to compensation. This is an approach taken by most of the major international consulting firms such as Ernst & Young or Towers Perrin. Microsoft's KM system, called ICE for Intellectual Capital Exchange, not only keeps track of who has shared information with whom, but requires the recipient of the information requested to grade the quality of the response so that information supplied is not only counted, but is graded and weighted as well. Microsoft does not heavily publicize the fact, but the information the system provides is used for evaluation and compensation purposes.

The story of the pharmaceutical sales rep who discovered an important lesson learned about the uptake of a new drug illustrates an important fact, that an organization's compensation structure may discourage knowledge sharing. Pharmaceutical sales reps are paid a very modest basic salary, and the bulk of their compensation comes from bonuses based on performance, related to their sales. Why should a sales rep share a lesson learned that makes her a star and gives her a bonus far in excess of the norm, when sharing the lesson learned is most likely to lead to her being returned to the middle of the pack and to receive a very average bonus? In fact, the sales rep had kept the lesson learned to herself until the KM team spotted her anomalous success and paid her a visit, and assured her that there would be something in it for her and that she would be feted for sharing and not pinged for being slow to share. The obvious conclusion is that if KM is to be taken seriously and implemented successfully, the organization's compensation practices need to be examined as to their impact on KM, and modified if necessary. This is, one must recognize, a tall order. Changing an organization's compensation scheme sets off waves, not just ripples, in all directions, and organizations tend to change or modify their compensation scheme very cautiously and slowly.

The other obvious answer to motivating participation is recognition. The World Bank, for example, has an annual "Information Fair" where employees who have made an important contribution are recognized, and where they are encouraged to mount a display describing their contribution and its impact [Srikantaiah, T., 2004].

7.7   KM METRIFICATION

The issue of measuring KM is a thorny one. The development of KM was very much derived from the recognition of the importance of intellectual capital and the desire to measure intellectual capital, however, as discussed above, IC is a very difficult thing to measure. Information or knowledge is abstruse and slippery. There are several sources of suggestions as to what KM related items can be measured and what indices can be created. Department of the Navy [2001] Metrics Guide for Knowledge Management Initiatives, developed by a team lead by Alex Bennett is particularly extensive. Although focusing on IC, Intellectual Capital, another source of extensive suggestions is a paper by Liebowitz and Ching [2000].

Metrics suggested for measuring KM generally fall into two classes, either measures of the use of KM components such as lessons learned databases and portals, expertise locators, and communities of practice, or they are general measures about the performance of the organization which would be useful to have, and in many cases would be maintained in any case, even if no KM initiative were under way. The latter measures are offered presumably on the supposition that if you can show that the organization is doing better, then you have a more persuasive case for KM. An argument for cause and effect is difficult to craft in this case because there may be many variables that account for an organization's success. Metrics for KM continue to be the 'holy grail' for KM advocates, and there is no one way to measure KM's correlation with mission accomplishment.

7.8   JUSTIFYING KM

The obvious justification for KM is to make the organization more effective and more competitive. All of the various homilies, now almost platitudes, apply, i.e., we are in the information age, work smarter, not harder, etc. The problem is that it is very difficult to document the effect of KM or of the larger information environment upon the results that matter to the organization. The reasons are that information or knowledge are very amorphous and difficult entities to quantify or measure, and because there are so many other complex and confounding variables that matter to the organization. Measures are fine as far as they go, and they are recommended if they make sense to a particular organizational context. A very good discussion of what measures can be used for justification is a paper by Powell, T. [2004a], entitled "Knowledge Return on Investment." He includes an ROI model and provides very useful commentary on what is practicable. Similarly, Lytras and Edvinsson [2008] in a chapter on how to justify and present an intellectual capital accounting, set forth an extensive menu of IC items that might be measured.

The consequence of that difficulty is that the technique that has occupied a sizeable chunk of the KM literature is that of storytelling [Brown et al., 2004, Denning, S., 2000]. Tell management a good true story of how knowledge sharing accomplished or at least enabled something of importance to the organization, or create a good "what if" story - 'what if the folks in the Chicago office received an RFP for ...from .... and were immediately made aware that last October the folks in the Singapore office responded successfully to a very similar RFP from ….' The power of the story can be as convincing as any metrics.

What else can one do? One can point out, as mentioned above that there is an extensive literature pointing out that organizational success, at least at the R&D level, is clearly associated with rich deep and open communication, and that the implementation of KM will foster such a knowledge sharing environment within the organization.

The pitch that KM will save time of key personnel and that, therefore, they will have more time to spend on other, presumably productive, activities is not a useful one. A well designed KM system can save the users' time, but there is an unrecognized danger in using that as a justification for investing in or improving a KM system. Of course, it makes sense to save the user's time, but the justification of the KM system ultimately has to be demonstrated by better decisions and improved performance.

The problem with using time saving as a justification is that even if the user's time is saved by a KM system, management probably won't see any observable difference in the user's behavior, and your justification could backfire on you. Why won't management see a difference?

The answer is a surprising one. Over the years, a number of studies of the practices of white-collar professional employees have been conducted, and findings are quite uniform and very corroborative. White-collar professional employees spend a very consistent 20–25% of their time information seeking [Griffiths, J., 1982, King et al., 1981, Nelke, M., 1999, Normier, B., 2001, Poppel, H., 1982, Roderer et al., 1983]. Workplace technology may have changed, but that 20-25% hasn't. This proportion is surprisingly independent of the apparent information intensity of the job domain. Line business managers and administrators generally spend as much of their time information seeking as do research scientists.

This commonality at first glance is surprising. Why does it emerge? The most likely answer is that there is some a sort of homeostasis, or perhaps more accurately, a satisfying mechanism at work. Knowledge workers whether managers or administrators or researchers need substantial information input to perform satisfactorily, but when the amount of time devoted to that function approaches 20%, then knowledge workers typically begin to satisfice. In other words, they begin to conclude that they have to get on with the rest of their job, and that if they have not already done so, they will soon run into diminishing returns in their information seeking, and that it is time to proceed based on the information they have.

The fact that there may be some sort of homeostatic phenomenon here is strongly hinted at in recent research on social learning by a team led by Kevin Laland [Macleod, M., 2010, Rendell et al., 2010]. Their research, consisted of a competition for a prize of 10,000 U.K. pounds for the best survival tactic for agents in a computer game, a 'survival of the fittest' game as it were. Each competitor was given 100 agents who would accumulate a repertoire of behavior over the agent's lifetime through learning. At each round, an agent would have three options, individual learning (acquiring a new behavior at random), social learning (acquiring a new behavior by observation), and exploitation (using a previously learned behavior). At each round, each agent had a 1 in 50 chance of dying, but was replaced by offspring from another agent. That parent agent could be from any competitor, and the chance of being chosen to be a parent was in proportion to each agent's success to date. Success was defined as an agent's accumulated score divided by the number of rounds it had survived. The offspring of a parent agent belonged to the parent agent's competitor. The better a strategy's performance, the greater the share of agents a competitor would accumulate, and the winner was the competitor whose strategy produced the most agents. The points to be observed is that the most successful strategies relied heavily on social learning and were characterized by spending between 10% to 20 % learning, either individual or social, and by spacing out that learning over the course of the game, rather than trying to load up at the beginning of the game, findings rather in line with the observations above.

The consequence of this phenomenon is that if a KM system allows a user to save time, that time will most likely be diverted into other information seeking behavior and the user will still spend the same 20–25% of their time information seeking. That is, of course, not bad; both the information need served by the KM system and the information need into which time was diverted may well be better served, with better decisions made and higher productivity resulting.

The point is that time saved will probably not be a very convincing indicator because it probably will not be demonstrable. If KM justifications are made to management on the grounds of time saved, then management will quite predictably and understandably expect to see that time saved deployed somewhere else like making sales calls or serving customers. When managers see that saved time simply directed to other information search time, if they notice any difference at all, and they probably won't, the behavior will look the same to them, and they will quite understandably feel that they have been misled and oversold.

7.8.1 SUCCESS

The most important justificatory technique however is success. Once a KM advocate is in a position to get underway, s/he can start with something small that will have visible results soon. Those results can justify further and larger steps. Note that it's important not just to have success with those first efforts, but visible success. Thus, the emphasis upon "pain points"[Durham, M., 2004, Hasanali, F., 2004, Short and Azzarello, 2004]. Look for a pain point, an area where the organization has a problem and where a success, even a small one, will be visible.

7.9 POST RETIREMENT KNOWLEDGE RETENTION

Awareness about the impending retirement of those in the baby boom bulge and the consequences it will have, demographic, financial, and health care related for example, has also created an awareness of the amount of knowledge that will, if traditional retirement practices are followed, be lost with those retirees. KM has, not illogically, been seen as the answer to that problem.

There are two KM practices that immediately come to mind and that have been promoted. One is a variation of the normally project-based after action report, that is to do an after action report based, as it were, on the employees career up to the time of retirement [Baxter, A., 2006].

The other technique is to keep the retiree active in the organization's CoPs [Hawamdeh and Refai, 2008]. The latter technique is far more likely to be useful the authors believe.

If retirees are encouraged to remain as active members of one or more communities of practice, then their knowledge is not lost, and the retiree may well be able to contribute valuable knowledge "Back when we introduced the model 812, we had similar teething problems, and the solution turned out to be …, and what we did was …." The organization has also gained, at minimal cost, important good will among their retirees and among the local communities where those retirees live.

The career debriefing preceding retirement, the data dump, and its possible conversion into lessons learned is not nearly so likely to be useful.
Why?

1. First, the likelihood that the information that will later prove useful, for example, the teething problems of the model 812, will come up in that "data dump" interview, even an extended one, before the employee retires is by no means certain, or even likely.

2. Second, the chance that someone in the organization will actually make the effort to look for relevant information in those "data dumps," even if converted into a "lesson learned," at the time it would be useful is even more minimal. Remember Mooers' law: (roughly recast) information will only be used if it is more trouble not to use it than it is to use it [Mooers, C., 1960].

3. Third, even if someone does search, the chance of finding the right information and making the connection is also rather small. One might find the problems with the model 812 by searching under, let us say, "lubrication problems," but the chances that the old timer will have used some other phraseology to describe the situation is probably equally high. A well developed taxonomy will help, but the complexities of language are enormous, and the retiree can refer to the incident in a myriad of ways.

4. Fourth, the response is very likely not as simple as "when we introduced the model 812, we had similar teething problems, and the solution turned out to be that …." It is more likely to be something like "back when we introduced the model 812, we had similar teething problems, and the situation here puts me in mind of that incident, and it occurs to me that what could be causing the problem is …, and a possible field fix might be …." In short, the great utility of the retiree's knowledge is likely to be not just the knowledge per se, but the ability to apply that knowledge to a new situation, and it is quite often the old timer who will have that ability, not the person who looks it up in the data dump.

5. Finally, the solution is likely to be found not directly in the information that the retiree leaves behind, but in new knowledge created by the interaction of the retiree with current employees. The retiree says "it occurs to me that …" and elicits a response something like "yes, but here …," a discussion unfolds, the retiree contributes some of the needed expertise, and a solution is generated.

The solution, therefore, is not likely to be the data dump and the attempt to stockpile knowledge in-house; the solution is far more likely to be opening up the organization's communities of practice to the alums and (frequently also) to selected members of the world outside the organization. Caterpillar Inc. is an organization that has been particularly pro-active in this use of KM for their "alums"[Stuedemann, R., 2004], and it was success in this venture that led them to go still further afield to dealers and repair facilities.

If an organization does intend to pursue the data dump/lessons learned approach, two recommendations are in order: first, the process should be formalized, and someone should be charged with the responsibility of conducting the interviews, and if at all possible, this should not be just an collateral duty; second, after the interview, both the interviewer and the interviewee should be charged with the tagging, and that tagging should be supported by a taxonomy. The data could benefit from being scrubbed, so that back channeling and redundancies can be reduced.

CHAPTER 8

Knowledge Management and Decision Making

One of the most beneficial applications of knowledge management is its role in advancing organization knowledge and the role knowledge plays in making decisions. A decision, according to Mintzberg et al. [1976], is "a specific commitment to action (usually a commitment of resources)" (p. 246) and a decision process is "a set of actions and dynamic factors that begins with the identification of a stimulus for action and ends with the specific commitment to action" (p. 246). Clearly, decision making is at the heart of effective and responsible management for individuals and organizations. Mintzberg, H. [1975], Mintzberg and Westley [2001] have studied and written eloquently on decision making in organizations since the 1970's. For the sake of understanding the way decisions are made, they have placed decision processes into three categories: the rational model (or thinking first), the creative discovery model (seeing first), and the action model (doing first).

The rational model, according to Mintzberg and Westley [2001], consists of a very orderly way to approach decisions and can also be called a "thinking first" model. The decision makers first define the problem and proceed from there. The next step is diagnosing the problem, and then imagining and designing possible solutions. Of course, the final step is to choose among the solutions and actually make the decision. This all sounds well and good, but in actual practice, decision makers are interrupted, events intrude, and emergencies present themselves. Research has shown that the decision process is a spiral one just as is the development of knowledge, that often keeps cycling back after interruptions [Nonaka and Takeuchi, 1995]. Decisions emerge, or sometimes, especially in times of stress and emergencies, they "erupt," so as a practical matter, decision makers need to expand their repertoire of decision making strategies.

An alternative model is what Mintzberg and Westley [2001] describe as a creative discovery model, or "seeing first," a method that relates to creative discovery. The steps in this model are preparation, incubation, illumination, and verification, steps that are again reminiscent of the SECI model of knowledge management already discussed [Nonaka and Takeuchi, 1995]. The model recognizes that managers often make decisions after groping around or working in a climate of information chaos. Because managers often have many different decisional roles such as negotiator, resource allocator, disturbance handler, entrepreneur, etc. [Mintzberg, H., 1975], they juggle many pieces of information along with what they know in their own internal "mental database." The information and their on-the-ground experience feed into what managers know and form a foundation for making decisions that often come as illuminations and depend on what is seen through insights as well as what is thought rationally.

In reality the creative discovery model is a bridge between the classic, rational model and something that Mintzberg and Westley [2001] call "doing first" – the way that pragmatists often make decisions. If the "thinking first" decision depends on verbal reasoning, and the "seeing first" model relies on the visual, the "doing first" decision style is more visceral, coming from intuition and instinct. As any number of analysts has written, the pressures placed upon organizational managers are varied and plentiful, and decisions are often made quickly, often by necessity [Gladwell, M., 2005, Gosling and Mintzberg, 2003, Mintzberg, H., 1975]. This natural flow of decision making can be effective if managers are committed to sharing knowledge through collaboration and by ongoing efforts to find knowledge that exists within an organization as well as the routine development of new knowledge for progress and innovation. As Mintzberg's research shows [Gosling and Mintzberg, 2003, Mintzberg, H., 1975], decision makers are seldom reflective, systematic planners. Instead, they depend on conversations and meetings to learn and form background for opinions. Because research has shown that managers often make decisions quickly and that they depend on interactions to build their own knowledge base, it is reasonable to assume that they participate in knowledge building through processes. Mackenzie (2005) has shown that managers in particular, as opposed to non-managers, tend to have a pattern of consciously and deliberately accumulating and retaining a "savings bank" of potentially useful knowledge.

CHAPTER 9

Social Network Analysis and KM

9.1 SOCIAL NETWORK ANALYSIS

Social Network Analysis has, or should have, an obvious application to KM. Allen's work, described under the section "Gatekeepers, Information Stars, & Boundary Spanners," was an early example of sociometric analysis and the use of sociograms, the technique that since it has become automated has now come to be called Social Network Analysis, or SNA. When person-to-person communication maps, or sociograms, had to be drawn manually, as Allen had to do, the technique was simply called sociometric analysis, but when digital email communications resulted in easily trackable person to person communications within and outside the organization, and software was created to recast that data into easily apprehensible graphic form, the result was such an "Aha, that's neat" breakthrough that a new name begged to be coined, and that new name was SNA, Social Network Analysis.

SNA provides the ability to easily identify the information stars within one's organization, without having to go through the labor-intensive techniques that Allen or Tushman had to use. The obvious fodder for SNA in this context is the email correspondence within the organization, but as social network tools or facilities such as Twitter and Facebook become more common in the business world, these are beginning to supply fodder as well. Those stars, thus identified, can then be connected to or assigned to key projects, and they can be identified to play key roles in Communities of Practice (CoPs). Again, though, like the overall lack of recognition of the importance of information mentioned above, there is still rather less recognition of the importance of SNA than there should be in the business and systems communities. A quick Google search of "Social Network Analysis" (as of September, 2010) results in about 11,400,000 hits, an impressively large number. But, a quick perusal of those hits shows comparatively little in the domain of business or systems, despite the popularity of SNA in the academic world, particularly in the fields of communication and sociology.

That is not to say that there has not been some important work in the area of business. Perhaps the most central and most often cited piece is that by Cross et al. [2002] on using SNA to support strategic collaboration. Their work emphasizes the importance of informal networks within the organization, and stresses that with increasingly de-layered and geographically dispersed organizations those informal communication networks are becoming more important and more vital. While their focus is top-down on strategically important groups such as top leadership networks and strategic business units, rather than a more bottom-up focus on projects, they do use "new product development teams" as an example of where SNA can be applied. Its publication in a prominent management journal, the California Management Review, has helped bring SNA to the attention of the management community. Specifically in the area of project management, Mead, S. [2001] introduces SNA and describes using SNA to analyze communications within a project team.

There are three major utilities for SNA in the context of KM:

1) SNA is a key tool with which to design and assemble a project team. Project teams should have an information star, a gatekeeper/boundary spanner as a key component of the team, and project teams should be put together keeping very consciously in mind that a known information star should be part of the team. It also allows for monitoring of the communication within a team, and can provide an early warning for teams or projects that do not appear to be communicating effectively.

   Give those information stars extra attention and extra support. Those stars have a leveraging effect. Support for them not only enhances their performance, *it enhances the performance of the project team*

2) SNA is a key tool with which to assemble, monitor, and coach an organization's Communities of Practice. Successful CoPs need Moderators, Thought Leaders, and Managers [Durham, M., 2004]. SNA helps identify candidates for all three roles, particularly Thought Leaders and Moderators.It also allows for the monitoring of CoPs, and can provide an early warning concerning CoPs that do not appear to be coalescing and communicating effectively, or that need attention or are in decline. Note that a CoP is not a forever institution; CoPs have a lifecycle and that includes dissolution when the time has come, and SNA can provide the clue as to when that time has come.

3) SNA is a key tool to diagnose communications emboli in your organization. The product development group in Parsippany, for example, should be in substantial communication with the engineering group in Terre Haute, but SNA can reveal that the communication is minimal, and can provide hints as to how to facilitate those communications and who might be the key players to utilize in enhancing those communications.

Table 9.1: Social Network Analysis Tools.

UCINET		`http://www.analytictech.com`
PAJEK	tutorial:	`http://iv.slis.indiana.edu/lm/lm-pajek.html`
	Downloadable from:	`http://pajek.imfm.si/doku.php`
NEGOPY		`http://www.sfu.ca/~richards/Pages/negopy.htm`

A good introduction (though last updated in 2001) to SNA with a list of packages is at

`http://lrs.ed.uiuc.edu/tse-portal/analysis/social-network-analysis`

Wikipedia has an extensive entry for "Social Network Analysis Software"

`http://en.wikipedia.org/wiki/Social_network_analysis_software`

Social Media

Analysis Tool:

CONTEXTMINER   `http://www.contextminer.org/`

CHAPTER 10

Implications for the Future

It is quite obvious to anyone who has studied the phenomenon of social media that the future of knowledge sharing is through the easy to use social software that is free or inexpensive and can reside not only on a PC but on any mobile device on in the "cloud." Often called Web 2.0 applications, these applications that ease communication and therefore knowledge sharing are blogs, wikis, chats, conversation bursts through Twitter and its ilk, and popular social websites, such as Facebook. As Frank Leistner says, "A knowledge flow lives and dies with participation" [Leistner, F., 2010, p. 150]. The millions who have adopted social networks and software demonstrate that knowledge sharing can be done in simple ways through easy to use tools. Organizations are encouraging participation by using blogs and wikis, often for internal knowledge sharing, and even Twitter and Facebook are showing up on corporate home pages. More important than the ease-of-use argument, social software often creates relationships and communities, thereby establishing trust, an important element to the willingness to share one's knowledge. Once again it's not about the technology, but about being able to connect with people and learn what they know.

There are drawbacks to social networking and social software to be sure. The communication is often very informal and sometimes incomplete, and archiving, organizing and retrieving the objects created are at the very nascent stages. We are just developing cyber ethnography tools and methodologies so that researchers can study how Web 2.0 applications can be used most effectively for knowledge management, knowledge sharing, and knowledge networking and flow. As Johnson states in his work on knowledge networking and such online web platforms as LinkedIn and MySpace, "…these sites offer considerable possibilities for sharing and transferring information that may be used to enhance creativity and innovation" [Johnson, J., 2009, pp. 107–108]. Even health clinicians, attorneys, and other professionals are coming to depend on and value the various connections that can be made through everyday ways to create new communities and learn from their members. Although knowledge management is not ABOUT the technology, the new interest in social networking may be responsible for a resurgence in attention being paid to knowledge exchange and knowledge sharing.

CHAPTER 11

Conclusion

This paper has reviewed the writing and scholarship on knowledge management in various fields in order to assist the reader in understanding the state of KM theory and its practical implications. The authors have written from the point of view of KM as a variety of processes instead of the management of knowledge primarily as management of thing(s). We have also considered KM in situ, in the workplace as it is practiced, and its relevance in day to day organizational management and decision making. And finally, the document has tried to delineate some of the possible future directions for KM.

Knowledge management has evolved since its introduction to information science, information systems, and other fields, from a perception of KM as a technological platform to today's understanding of KM as communication, information, and organizational practices within intentional communities and social groups. Because of the adoption of knowledge management practices in organizations, large and small, and the ongoing interest in KM by scholars and practitioners alike, knowledge management is likely to be a fruitful area for research by all those in the information sciences, computer science, business administration, communication, and organizational science.

APPENDIX A

KM Websites

1. American Productivity and Quality Center: http://www.apqc.org. The American Productivity and Quality Center was a competitive response to Japanese technological and managerial advances in the 1970s to make US companies more competitive by sharing best practices. Not surprisingly, it latched onto KM enthusiastically and has a great deal of KM related material. Just key in "KM" in the search box to find it.

2. Brint: http://www.brint.com. Brint has been around for a while, covers many topics, including much on KM, but getting very cluttered and overly busy and distracting.

3. CIO: http://www.cio.com. Key in KM in the search box. There are a number of good articles here, but the site is cluttered, busy, and difficult to navigate, and what you get isn't always what you think you clicked on.

4. David Skyrme Associates: www.skyrme.com/insights/22km.htm. David Skryme is a pioneer in KM, and this site has some very good insights and practical advice.

5. Gurteen Knowledge: http://www.gurteen.com/. David Gurteen is one of the most intelligent and active knowledge management evangelists in the world. He is a clear thinker and a generous speaker. His website offers a wealth of information about KM and new media. He is the founder of the Knowledge Café process as explained on the site.

6. ICASIT: http://www.icasit.org/km/resources/kmcases.htm. A very good source for KM Case Studies (still a surprisingly scarce commodity).

7. KMCI Press: http://www.kmci.org. KMCI Press is an imprint of Butterworth-Heinemann, a division of Elsevier Science and Technology Books. The site directs you to not only their publications, but to a number of other sites as well.

8. KM Network: http://kmnetwork.com/. Given that the main purpose of a KM portal is to provide a comprehensive array of information about the topic, www.kmnetwork.com lived up to this criterion. It has an extensive range of KM-related resources.

9. KM World: http://www.kmworld.com. KM world is one of the best sources for news about software, conferences and other developments related to knowledge management.

10. Knowledge Board: Your Global Community: http://www.knowledgeboard.com/. This portal has a library of over 300 articles/web portals. It seems to be updated quite regularly, and it has active forums/discussion boards.

11. Knowledge Enhancement Resource Center: `http://www.kmresource.com`. This site focuses primarily on KM products, but it also has a great deal of KM information, including a number of classic articles, freely available, plus introductory articles, a KM glossary and other goodies. The too busy presentation is annoying, but good stuff behind it.

12. Straits Knowledge: `http://www.straitsknowledge.com`. The leading KM consultancy in Asia, with an accurate products to guide KM implementation. There are also a number of free KM publications on topics such as "Key Performance Indicators."

13. Strategic and Competitive Intelligence Professionals: `http://www.scip.org`. SCIP, Strategic and Competitive Intelligence Professionals, formerly the Society of Competitive Intelligence Professionals has interests that overlap greatly with KM. You need to be a member to utilize the website, however.

14. TFPL: `http://www.tfpl.com`. TFPL is a London based consultancy with expertise in KM and a number of useful documents on topics such as communities of practice freely downloadable.

15. Tom Davenport: `http://www.tomdavenport.com/`. Davenport is well known in the KM field. He's moved on to Analytics, but his website and blog are filled with interesting musings, insights and information that could generally fall under a KM rubric.

16. Working Knowledge: `http://hbswk.hbs.edu/`. Harvard University shows how knowledge management can be put to work with its website devoted to faculty research.

17. World Bank: `http://www.worldbank.org`. A key player in developing and promoting KM, their website contains a great deal of KM information.

Bibliography

Abrahamson, E. (1996) Management fashion. *Academy of Management Review*, 21(1), 254–285. DOI: 10.2307/258636

Abrahamson, E. and Fairchild, G. (1999) Management fashion: lifecycles, triggers, and collective learning processes. *Administrative Science Quarterly*, 44, 708–740. DOI: 10.2307/2667053

Ackerman, M., Pipek, V., and Wulf, V. (Eds.), (2003) *Sharing expertise: Beyond knowledge management*. Cambridge: The MIT Press. 35

Allen, T. and Cohen, S. (1969) Information Flow in Research and Development Laboratories. *Administrative Science Quarterly*, 14(1), 12–19. DOI: 10.2307/2391357 23

Allen, T. (1977) *Managing the Flow of Technology: Technology Transfer and the Dissemination of Technological Information Within the R&D Organization*. Cambridge, MA: MIT Press. 13, 23

Argyris, C. and Schon, D. (1978) *Organizational learning: A theory of action perspective*. Reading, MA: Addison Wesley. 19, 35

Argyris, C. and Schon, D. (1996) *Organizational learning II: Theory, method and practice*. Reading, MA: Addison Wesley. 35

Barth, S. (2004) Integrating Knowledge management and Competitive Intelligence: Integrating Offense and Defense. In M.E.D. Koenig and T.K. Srikantaiah, (Eds.), *Knowledge Management Lessons Learned: What Works and What Doesn't*. (pp. 461–481). Medford, NJ: Information Today, for the American Society for Information Science and Technology. 50

Baumard, P. (1999) *Tacit knowledge in organizations*. London: Sage Publications. 19, 35

Baxter, A. (2006) Keeping the Know- How of a Retiring Generation. *Financial Times*. 54

Belkin, N., Oddy, R., and Brooks, H. (1982) ASK for information retrieval, Part 1. Background and theory. *Journal of Documentation*, 38, 61–71. DOI: 10.1108/eb026722 32

Bernstein, L., Siegel, E., and Goldstein, C. (1980) The Hepatitis Knowledge Base: A Prototype Information Transfer System, *Annals of Internal Medicine*, Vol. 93, Number 1, Part 2, 169–181. 40

Blackler, F. (1995) Knowledge, knowledge work and organizations: An overview and interpretation. *Organization Studies*, 16, 1021–1046. DOI: 10.1177/017084069501600605 20, 29

Boahene, M. and Ditsa, G. (2003) Conceptual confusions in knowledge management and knowledge management systems: Clarifications for better KMS development. In E. Coakes, (Ed.), *Knowledge management: Current issues and challenges.* (pp. 12–24). Hershey, PA: IRM Press. 31

Bouthillier, F., and Shearer, K. (2002) Understanding knowledge management and information management: The need for an empirical perspective. *Information Research,* 8(1), Article 140. Retrieved August 21, 2006, from `http://informationr.net/ir/8--1/paper140.html`.

Brown, J. and Duguid, P. (2000a) *The Social Life of Information.* Boston, MA: Harvard Business School Press. 26

Brown, J. and Duguid, P. (2000b) Balancing Act: How to Capture Knowledge Without Killing It. *Harvard Business Review,* 78(3), 73–80.

Brown, J., Denning, S., Groh, K., and Prusak, L. (2004) *Storytelling in Organizations: Why Storytelling is Transforming 21st Century Organizations and Management.* Burlington, MA: Butterworth-Heinemann. 52

Buckman, R. (2004) *Building a Knowledge-Driven Organization.* New York, NY: Mcgraw- Hill. 5

Burk, C. and Horton. F. (1988) *InfoMap: A Complete Guide to Discovering Corporate Information Resources.* Englewood Cliffs: Prentice Hall. 37

Chait, L. (2008) Sharing Knowledge Problems, Root Causes and Solutions. In M.E.D. Koenig and T. K. Srikantaiah (Eds.), *Knowledge Management In Practice, Connections and Context.* (pp. 277–288). Medford NJ: Information Today, for The American Society for Information Science and Technology. 36

Chatzkel, J. (2000) Conversation with Hubert St. Onge, Senior Vice President for Strategic Capabilities at Clarica Life Insurance Company, interview, Journal of Intellectual Capital. DOI: 10.1108/14691930010371645 5

Chiu, C., Hsu, M., and Wang, E. (2006) Understanding knowledge sharing in virtual communities: An integration of social capital and social cognitive theories. *Decision Support Systems,* 42(3), 1872–1888. DOI: 10.1016/j.dss.2006.04.001 49

Choo, C. W. (2002) Sensemaking, knowledge creation, and decision-making: Organizational knowing as emergent strategy. In C. W. Choo and N. Bontis (Eds.), *The strategic management of intellectual capital and organizational knowledge,* (pp. 79–88). New York, NY: Oxford University Press. 33

Combs, R. and Moorhead, J. (1993) *The Competitive Intelligence Handbook.* Metuchen, NJ: The Scarecrow Press. 50

Cross, R., Borgatti, S., and Parker, A. (2002) Making Invisible Work Visible: Using Social Network Analysis to Support Strategic Collaboration. *California Management Review*, 44(2), 25–46. 59

Davenport, T. and Prusak, L. (1998a) *Working knowledge: How organizations manage what they know.* Boston, MA: Harvard Business School Press. 6, 26, 36, 47

Davenport, T. (1998b) Successful Knowledge Management Projects. *Sloan Management Review*, 39(1), 43–57.

Davenport, E. (2004) Interpersonal Knowledge and Organizational Foresight: The Case of Online Partnership in Micro-Organizations.In M.E.D. Koenig and T. K. Srikantaiah (Eds.), *Knowledge Management: Lessons Learned: What Works and What Doesn't.* (pp. 379 - 397). Medford, NJ: Information Today, for The American Society for Information Science and Technology. 49

DeLong, D.W. (2004). *Lost Knowledge: Confronting the Threat of the Aging Workforce.* NY: Oxford University Press, USA. 8

Denning, S. (2000) The Springboard: How Storytelling Ignites Action in Knowledge-Era Organizations, Burlington, Mass., Butterworth-Heinemann. 52

Department of the Navy (2001) Department of the Navy , Chief Information Officer. Available from www.don-imit.navy.nil 52

Desouza, K. C., Awazu, Y. and Wan, Y. (2006) Factors governing the consumption of explicit knowledge. *Journal of the American Society for Information Science and Technology*, 57(1), 36–43. DOI: 10.1002/asi.20250 32

Dimitrova, D., Garton, L., Gulia, M., and Haythornthwaite, C. (1996) Computer networks as social networks: Collaborative work, telework, and virtual community. *Annual Review of Sociology*, 22, 213–238. DOI: 10.1146/annurev.soc.22.1.213

Drucker, P. F. (1993) *Post-Capitalist Society.* New York, NY: Harper Business division of Harper Collins. 4

Drucker, P. F. (2005) Managing oneself. *Harvard Business Review*, 83(1), 100–109.

Durham, M. (2004) Three Critical Roles for Knowledge Management Workspaces. In M.E.D. Koenig and T. K. Srikantaiah (Eds.), *Knowledge Management: Lessons Learned: What Works and What Doesn't.* (pp. 23–36). Medford NJ: Information Today, for The American Society for Information Science and Technology. 43, 54, 60

Edvinsson, L. (1994) Visualizing Intellectual Capital in Skandia. *Skandia's 1994 Annual Report.* Stockholm: Skandia AFS. 5

Edvinsson, L. (1997a) Developing Intellectual Capital at Skandia. *Long Range Planning* 30(3), 366–373. DOI: 10.1016/S0024-6301(97)00016-2

Edvinsson, L. and Malone, M. (1997b) *Intellectual Capital, Realizing your Company's True value by Finding Its Hidden Brainpower.* New York, NY: Harper Business/Harper Collins. 5

Ehrlich, K. (2003) Locating Expertise: Design issues for an expertise locator system. In M. Ackerman, V. Pipek and V. Wulf (Eds.), *Sharing Expertise: Beyond knowledge management* (pp. 137–158). Cambridge: The MIT Press. 35

Engeström, Y. (1999) Expansive visibilization of work: An activity-theoretical perspective. *Computer supported cooperative work*, 8, 63–93. 20, 29

Fishkin, J. (2000) *KM and OL from an IT Perspective*, a presentation to the 2001 Conference Board conference: The 2001 Knowledge Management/Organizational Learning Conference: Convergence, Application, and Infrastructure, 3–4 May, New York, NY.

Friedman, T. L. (2005) *The world is flat: A brief history of the twenty-first century.* New York, NY: Farrar, Straus, and Giroux. 37

Frost, P. and Whitley, R. (1971) Communication Patterns in a Research Laboratory. *R and D Management*, 1(2), 71–79. 24

Gladwell, M. (2005) *Blink: The power of thinking without thinking.* New York, NY: Little, Brown & Co. 58

Goldhar, J., Bragaw, L., and Schwartz, J. (1976) Information Flows, Management Styles, and Technological Innovation. *IEEE Transactions on Engineering Management, EM-23* (1), 51–61. 13, 26

Gosling, J. and Mintzberg, H. (2003) The five minds of a manager. *Harvard Business Review*, 81(11), 54–63. 58

Grant, R. M. (1996) Toward a knowledge-based theory of the firm. *Strategic Management Journal*, 17, 109–122. 19

Griffiths, J. (1982) The Value of Information and Related Systems. Products and Services. In: Martha E. Williams (Ed.). *Annual Review of Information Science and Technology.* (pp. 269–284). White Plains, NY: Knowledge Industry Publication, Inc. for the American Society for Information science. 53

Harvard Business Review. (1998) *Harvard Business Review on knowledge management.* Boston, MA: Harvard Business School Publishing. 35

Hasan, H. (2003) *An activity-based model of collective knowledge.* Paper presented at the 36th Hawaii International Conference on System Sciences (HICSS'03), Hawaii. DOI: 10.1109/HICSS.2003.1174230 29

Hasanali, F. (2004) Critical Success Factors of Knowledge Management. In M.E. D Koenig and T. K. Srikantaiah (Eds.) *Knowledge management, Lessons learned: What works and what doesn't.* (pp. 55–69). Medford, NJ: Information Today. 54

Hawamdeh, S. and Refai, H. (2008) Video Management and the Transfer of Knowledge, through Audiovisual Material. In M.E.D. Koenig and T. K. Srikantaiah (Eds.), *Knowledge Management In Practice, Connections and Context.* (pp. 223 - 238). Medford NJ: Information Today, for The American Society for Information Science and Technology. 45, 55

Hibbard, J. (1997) Knowing What We Know. *Information Week*, (20 October), 46–64. 4

Hicks, B. J., Culley, S. J., Allen, R. D., and Mullineux, G. (2002) A framework for the requirements of capturing, storing and reusing information and knowledge in engineering design. *International Journal of Information Management*, 22, 263–280. DOI: 10.1016/S0268-4012(02)00012-9 32

Hirschheim, R., Klein, H. K., and Lyytinen, K. (1995) *Information systems development and data modeling: Conceptual and philosophical foundations.* New York, NY: Cambridge University Press. 31

Hislop, D. (2002) Mission impossible? Communicating and sharing knowledge via information technology. *Journal of Information Technology*, 17, 165–177. DOI: 10.1080/02683960210161230 27

Horton, F. (1988a) Mapping Corporate Information Resources. *International Journal of Information Management*, 8(4), 249–259 DOI: 10.1016/0268-4012(88)90032-1

Horton, F. (1988b) Mapping Corporate Information Resources. *International Journal of Information Management*, 9(1), 19–24 DOI: 10.1016/0268-4012(89)90033-9

Horton, F. (1989) Mapping Corporate Information Resources. *International Journal of Information Management*, 9(2), 91–95. DOI: 10.1016/0268-4012(89)90018-2

Hsu, M., Ju, T., Yen, C., and Chang, C. (2007) Knowledge sharing behavior in virtual communities: The relationship between trust, self-efficacy, and outcome expectations. *International Journal of Human-Computer Studies*, 65(2), 153–169. DOI: 10.1016/j.ijhcs.2006.09.003 49

Inkpen, A., and Tsang, E. (2005) Social capital, networks, and knowledge transfer. *Academy of Management Review*, 30(1), 146–165. DOI: 10.2307/20159100 49

Isabel: Knowledge, Diagnosis,Treatment. `http://www.isabelhealthcare.com/home/default`

Johnson, J. D. (2009) *Managing knowledge networks.* Cambridge: Cambridge University Press. 63

Kaplan, R.S. and Norton, D.P. (1992) The Balanced Scorecard - Measures that Drive Performance. *Harvard Business Review*, 70(1), 71–91.

Killing off the American future. (2006) *The New York Times*, pp. A14. 37

King, D., McDonald, D., and Roderer, N. (1981) *Scientific Journals in the United States: Their Production, Use, and Economics.* Stroudsburg, PA: Hutchinson Ross Publishing Co. 53

Knowledge Management Solutions, Inc. Next Generation Human Capital Management Solutions. Retrieved October 7, 2005, from `http://www.kmsi.us/`

Koenig, M.E.D. (1990) Information Services and Downstream Productivity. In Martha E. Williams (Ed.), *Annual Review of Information Science and Technology: Volume 25,* (pp. 55–56). New York, NY: Elseview Science Publishers for the American Society for Information Science. 13

Koenig, M.E.D. (1992a) The Information Environment and the Productivity of Research. In H. Collier (Ed.), *Recent Advances in Chemical Information*, (pp. 133–143). London: Royal Society of Chemistry. 2, 13, 25

Koenig, M.E.D. (1992b) Entering Stage III-The Convergence of the Stage Hypotheses. *Journal of the American Society for Information Science*, 43(3), 204–207.
DOI: 10.1002/(SICI)1097-4571(199204)43:3%3C204::AID-ASI2%3E3.0.CO;2-F

Koenig, M.E.D. (1996) Intellectual Capital and Knowledge Management. *IFLA Journal, 22.04*, 299–301. 5

Koenig, M.E.D. (1998) From Intellectual Capital to Knowledge Management: What Are They Talking About?. *INSPEL,32.04*, 222–233. 5

Koenig, M.E.D. (2000a) The Evolution of Knowledge Management. In T. K. Srikantaiah and M.E.D. Koenig, *Knowledge Management for the Information Professional.* (pp. 23–36). Medford, NJ: Information Today for the American Society for Information Science. 4, 12

Koenig, M.E.D. (2000b) Information Driven Management, The New But Little Perceived, Business Zeitgeist. *Libri*, 50(3), 174–190.

Koenig, M.E.D. (2005) KM moves beyond the organization: An opportunity for librarians. Knowledge Management Section. World Library and Information Conference, sponsored by the International Federation of Library Associations and Institutions. Retrieved September 13, 2005. from `http://www.ifla.org/IV/ifla71/papers/123eKoenig.pdf` 13, 15

Koenig, M.E.D. and Neveroski, K. (2008) The Origins and Development of Knowledge Management. *Journal of Information and Knowledge Management*, 7(4), 243–254. DOI: 10.1142/S0219649208002111

Kwan, M. and Balasubramanian, P. (2003) KnowledgeScope: Managing knowledge in context. *Decision Support Systems*, 35, 467–486. 29

Koenig, M.E.D. (2009) Productivity Impacts of Libraries and Information Services. In: Marcia Bates and Mary Niles Maack (Eds.), *Encyclopedia of Library and Information Sciences*. CRC Press, Taylor & Francis, 2009. 26

KPMG Consulting. (2000) *Knowledge Management Research Report 2000*. New York, NY: KPMG Consulting. 46

Learn, L. L. (2002) The Internet as a mechanism to enhance content.In R.F. Bellaver and J.M. Lusa (Eds.), *Knowledge management strategy and technology*. (pp. 151–168). Boston, MA: Artech House. 35, 36

Leistner, F. (2010) *Mastering organizational knowledge flow: How to make knowledge sharing Work*. Hoboken, NJ: Wiley. 3, 4, 63

Liebowitz, J. and Ching, Y. Suen. (2000) Developing Knowledge Management Metrics for Measuring Intellectual Capital. *Journal of Intellectual Capital*, 1(1), 54–67. DOI: 10.1108/14691930010324160 52

Lin, M., Hung, S., and Chen, C. (2009) Fostering the determinants of knowledge sharing in professional virtual communities. *Computers in Human Behavior*, 25(4), 929–939. DOI: 10.1016/j.chb.2009.03.008 49

Lytras, M. D., Poulodi, N., and Poulymenakou, A., (2000) Knowledge Management Convergence: Expanding Learning Frontiers. *Journal of Knowledge Management*, 6(1), 40–51. DOI: 10.1108/13673270210417682

Lytras, M., Edvinsson, L, and Ordonez de Pablos, P. (2008) In M. Lytras, R. Meir, R. Maier, and A. Naeve (Eds.), *Knowledge Management Strategies: A Handbook of Applied Technologies*. (pp. 63–90). Hershey, PA: IGI Publishing. 52

Macleod, M. (2010) You Are What You Copy. *New Scientist*, pp.40–43. 53

Maag, G. and Flint, J. (2004) The Role of Corporate Intelligence Gathering in the Modern Business Decision-Making Process. In M.E.D. Koenig and T.K. Srikantaiah, (Eds.), *Knowledge Management Lessons Learned: What Works and What Doesn't*. (pp. 403–440). Medford, NJ: Information Today, for the American Society for Information Science and Technology. 50

Maier, R. (2002) *Knowledge management systems: Information and communication technologies for knowledge management*. Berlin: Springer-Verlag. 27

Malhotra, Y. (2000) Why Knowledge Management Systems Fail: Enablers and Constraints of Knowledge Management in Human Enterprises. In M.E.D. Koenig and T. K. Srikantaiah (Eds.), *Knowledge Management: Lessons Learned: What Works and What Doesn't*. (pp. 23–36). Medford NJ: Information Today, for The American Society for Information Science and Technology.

Malhotra, Y. (2004) Why Knowledge Management Systems Fail: Enablers and Constraints of Knowledge Management in Human Organizations.In M.E.D. Koenig and T. K. Srikantaiah (Eds.), *Knowledge Management: Lessons Learned: What Works and What Doesn't.* Medford NJ: Information Today, for The American Society for Information Science and Technology. 49

Managers Look to the Social Network to Seek Information.(2005). *Information Research, An International Electronic Journal,* 10(2), Paper 216. ISSN 1368–1613. Available from: `http://informationr.net/ir/10--2/paper216.html`

Marchand, D. (1985) Information Management: Strategies and Tools in Transition. *Information Management Review,* 1(1), 27–37. 2

Markus, M. L. (2001) Toward a theory of knowledge reuse: Types of knowledge reuse situations and factors in reuse success. *Journal of Management Information Systems,* 18(1), 57–93. 29

Mattox, D., Maybury, M., and Morey, D. (1999) Enterprise Expert and Knowledge Discovery. In *Proceedings of the 8th International Conference on Human Computer Interaction, Volume II (Communication, Cooperation, and Application Design),* pp. 303–313. 42

Mazzie, M. (2003) Personal Communication. 41

Maybury, M. (2003) Knowledge Management at the Mitre Corporation. In M. Rao (Ed.), *Knowledge Management in IT/Consulting Companies.* New York, NY: McGraw-Hill. 42

McInerney, C.R. (2002) Knowledge management and the dynamic nature of knowledge. *Journal of the American Society for Information Science and Technology,* 53(12), 1009–1018. DOI: 10.1002/asi.10109 1, 35

McInerney, C.R. (2000) *Providing data, information, and knowledge to the virtual office: Organizational support for remote workers.* Washington, DC: Special Libraries Association. 43

McKnight, D., Cummings, L., and Chervany, N. (1998) Initial trust formation in new organizational relationships. *Academy of Management Review,* 23(3), 473–49. DOI: 10.2307/259290 48

McMahon, C., Lowe, A., and Culley, S. (2004) Knowledge management in engineering design: Personalization and codification. *Journal of Engineering Design,* 15(4), 307–325. DOI: 10.1080/09544820410001697154 32, 33

Mead, S. (2001) Using Social Network Analysis to Visualize Project Teams. *Project Management Journal,* 32(4), 32–38. 60

Mintzberg, H. (1975) The manager's job: Folklore and fact. *Harvard Business Review,* 49–61. 57, 58

Mintzberg, H., Raisinghani, D. and Theoret, A. (1976) The structure of "unstructured" decision processes. *Administrative Science Quarterly,* 21, 246–275. DOI: 10.2307/2392045 57

Mintzberg, H. and Westley, F. (2001) Decision making: It's not what you think. *MIT Sloan Management Review,* 42(3), 89–93. 57, 58

Mondschein, L. (1990) SDI Use and Productivity in the Corporate Research Environment. *Special Libraries, 1990 Fall,* 81(4), 265–279. 13, 24

Mooers, C. (1960) Moor's Law: or, why some retrieval systems are used and others are not. *American Documentation,* 11(3), 204. DOI: 10.1002/bult.37 55

Morville, P. (2005) *Ambient findability.* Sebastopol, CA: O'Reilly Media. 36

Moulton, L. (2008) Knowledge Audits: Establishing a Context for Leveraging Knowledge. In M.E.D. Koenig and T. K. Srikantaiah (Eds.), *Knowledge Management: Lessons Learned: What Works and What Doesn't,* (pp. 79–93) Medford NJ: Information Today, for The American Society for Information Science and Technology. 38

Murtomaa, E. (1998) The Impact of the Functional Requirements for Bibliographic Records Recommendations on the ISBDER.64th Annual International Federation of Library Associations and Institutions General Meeting. Retrieved on April 23, 2004, from `http://www.ifla.org/IV/ifla64/065--74e.htm` DOI: 10.1300/J104v28n01_04

Nelke, M. (1991) Knowledge Management in Swedish Corporations. In M. E. D Koenig and T. K. Srikantaiah (Eds.), *Knowledge Management for the Information Professional.* Medford, NJ, Information Today for the American Society for Information Science.

Nelke, M. (1999) The Role of the Corporate Library in the Knowledge Management Process. *Information Services and Use,* 19(1), 49–54. 53

Nick, M., Althoff, K-D., and Tautz, C. (2001) Systematic maintenance of corporate experience repositories. *Computational Intelligence,* 17(2), 364–386. DOI: 10.1111/0824-7935.00150 32

Nissen, M., Kamel, M., and Sengupta, K. (2000) Integrated Analysis and Design of Knowledge Systems and Processes. *Information Research Management Journal,* 13(1), 24–43.

Normier, B. (2001)report of a study conducted by Lazard Freres and Co. LLC in 2001, presentation on "Natural Language Advances" at the KM World Conference, Santa Clara, California, 31 October 2001. 53

Nonaka, I. and Takeuchi, H. (1995) *The knowledge creating company*: *How Japanese Companies Create the Dynamics of Innovation.* New York: Oxford University Press. 6, 20, 45, 57

Nonaka, I. (1994) A dynamic theory of organizational knowledge creation. *Organizational Science,* 5(1), 14–37. DOI: 10.1287/orsc.5.1.14 20

O'Dell, C. and Jackson C. (1998) If only we knew what we know: Identification and transfer of internal best practices. *California Management Review*, 40(3), 154–174. DOI: 10.1225/CMR114 6, 31

Orpen, C. (1985) The Effect of Managerial Distribution or Scientific and Technical Information on Company Performance. *R&D Management*, 15(4), 305–308. DOI: 10.1111/j.1467-9310.1985.tb00041.x 13, 26

Peters, T. (1992) *Liberation Management: Necessary Disorganization for the Nanosecond Nineties*. New York, NY: Alfred A. Knopf. 4

Ponzi, L., and Koenig, M.E.D. (2002) Knowledge Management: Another Management Fad?",*Information Research*, 8(1). Retrieved from `http://informationr.net/ir/8--1/paper145.html` 15

Ponzi, L. (2004) Knowledge management: Birth of a discipline. In M.E. D Koenig and T. K. Srikantaiah (Eds.) *Knowledge management, Lessons learned: What works and what doesn't*, (pp. 9–26). Medford, NJ: Information Today. 15

Poppel, H. (1982) Who needs the Office of the Future?. *Harvard Business Review*, 60(6), 146–155 1982. 53

Powell, T. (2001) The Knowledge Value Chain (KVC): How to fix it when it breaks. In Martha Williams (Ed.), *Proceedings of the 22nd National Online Meeting*. Medford, NJ: Information Today. 35

Powell, T. (2004) A). Knowledge Return on Investment. In M.E.D. Koenig and T. K. Srikantaiah (Eds.), *Knowledge Management: Lessons Learned: What Works and What Doesn't*, (pp. 125–139) Medford NJ: Information Today, for The American Society for Information Science and Technology. 52

Powell, T. (2004) The Knowledge matrix, A Proposed Taxonomy for Enterprise Knowledge. In M.E.D. Koenig and T. K. Srikantaiah (Eds.), *Knowledge Management: Lessons Learned: What Works and What Doesn't*, (pp. 225 – 237). Medford NJ: Information Today, for The American Society for Information Science and Technology. 39

Powell, T. (2001) The Knowledge Value Chain (KVC): How to fix it when it breaks. In Martha Williams (Ed.), *Proceedings of the 22nd National Online Meeting*. Medford, NJ: Information Today.

Prusak, L. (1998) *Managing Principal, IBM Global Services, Consulting Group Presentation to the Conference Board*. Paper presented at the 1998 Conference on Knowledge Management and Organizational Learning, Chicago, 16 April. 48

Prusak, L. (1999) Where did Knowledge Management Come From?. *Knowledge Directions*, 1(1), 90–96. DOI: 10.1147/sj.404.01002 2

Prusak, L. (2001) Presentation at the International Federation of Library Associations Conference, Boston, MA. 45

Prusak, L. (2004) Personal Communication. 41

Quigley, E. J. and Debons, A. (1999) Interrogative theory of information and knowledge. In *Proceedings of the 1999 ACM SIGCPR conference on Computer Personnel Research*, pp 4–10, New Orleans, Louisiana, United States. New York: ACM Press. DOI: 10.1145/299513.299602 31

Rendell, L., Boyd, R., Cownden, D., Enquist, N., Eriksson, K., Feldman, M., Fogarty, L., Ghirlanda, S., Lillicrap, T., and Laland, K. (2010) Why Copy Others? Insights from the Social Learning Strategies Tournament. *Science,* Vol. 328, 5975, 208 – 213. DOI: 10.1126/science.1184719 53

Renzl, B. (2008) Trust in management and knowledge sharing: The mediating effects of fear and knowledge documentation. *Omega*, 36(2), 206–220. DOI: 10.1016/j.omega.2006.06.005 49

Richter, H., Abowd, G., Miller, C., and Funk, H. (2004) Tagging knowedge acquisition sessions to facilitate knowledge traceability. *International Journal of Software Engineering and Knowledge Engineering,* 14(1), 3–19. DOI: 10.1142/S0218194004001543 32

Ridings, C., Gefen, D., and Arinze, B. (2002) Some antecedents and effects of trust in virtual communities. *The Journal of Strategic Information Systems,* 11(3–4), 271-295. DOI: 10.1016/S0963-8687(02)00021-5 48

Robins, J. (2002) Affording a place: The role of persistent structures in social navigation. *Information Research,* 7(3). Article 131. Retrieved August 21, 2006, from `http://informationr.net/ir/7--3/paper131.html` 27

Roderer, N., King, D., Brouard, S. (1983) *The Use and Value of Defense Technical Information Center Products and Services.* Rockville, MD: King Research, Inc. 53

Rosenfeld, L. and Morville, P. (2002) *Information architecture.* Sebastopol, CA: O'Reilly Media. 36

Rotter, J. (1967) A new scale for the measurement of interpersonal trust. *Journal of Personality*, 35(4), 651–665. DOI: 10.1111/j.1467-6494.1967.tb01454.x 49

Rotter, J. (1971) Generalized expectancies for interpersonal trust. *American Psychologist*, 26(5), 443–452. DOI: 10.1037/h0031464 49

Ruggles, R. (1998) The State of the Nation: Knowledge Management in Practice. *California Management Review*, 40(Spring), 80–89.

Ruhleder, K. (2002) Understanding on-line community: The affordances of virtual space. *Information Research,* 7(3). Article 132. Retrieved August 21, 2006, from `http://informationr.net/ir/7--3/paper132.html` 27

Shelfer, K. (2004) Using Competitive Intelligence to Improve Knowledge Management. In M.E.D. Koenig and T.K. Srikantaiah, (Eds.), *Knowledge Management Lessons Learned: What Works and What Doesn't.* (pp. 441–460). Medford, NJ: Information Today, for the American Society for Information Science and Technology. 50

Searle, J. (1969) *Speech acts.* Cambridge, England: Cambridge University Press. 31

Selvin, A. and Buckingham Shum, S. (2002) Rapid knowledge construction: A case study in corporate contingency planning using collaborative hypermedia. *Knowledge and Process Management,* 9(2), 119–128. DOI: 10.1002/kpm.140 32

Senge, P.(1990) *The Fifth Discipline: The Art and Practice of the Learning Organization.* New York, NY: Doubleday Currency. 6, 19

Short, T. and Azzarello, Richard C. (2004) Knowledge Management in Action, Nine Lessons Learned. In M.E. D Koenig and T. K. Srikantaiah (Eds.) *Knowledge management, Lessons learned: What works and what doesn't.* (pp. 31–53). Medford, NJ: Information Today. 54

Srikantaiah, T. K. and Koenig, M.E.D. (Eds.). (2000) *Knowledge Management for The Information Professional.* Medford, NJ: Information Today for the American Society for Information Science.

Srikantaiah, T. K. (2004) Historical and Contemporary Perspectives on Knowledge Management: - and a Look at the Knowledge Sharing Initiative at the World Bank. In M.E.D. Koenig and T. K. Srikantaiah (Eds.), *Knowledge Management: Lessons Learned: What Works and What Doesn't.* (pp. 361 - 378). Medford NJ: Information Today, for The American Society for Information Science and Technology. 7, 49, 51

Stuedemann, R. (2007) Personal Communication. 7

Stewart, T. (1994) Your Company's Most Valuable Asset: Intellectual Capital. *Fortune,*pp. 68–74. 5

Stuedemann, R. (2004) Knowledge management at Caterpillar. In M.E.D. Koenig and T. K. Srikantaiah (Eds.), *Knowledge Management: Lessons Learned: What Works and What Doesn't.* (pp. 353 - 359). Medford NJ: Information Today, for The American Society for Information Science and Technology. 8, 56

Sullivan, Patrick H. (2000) *Value-Driven Intellectual Capital: How to Convert Intangible Corporate Assets into Market Value.* New York, NY: Wiley. 5

Sutton, M. (2008) *Examination of the Historical Sensemaking Processes Representing the Development of Knowledge Management Programs in Universities.* ,the ICKM (International Conference on Knowledge Management), Columbus Ohio, October 23. 1

Sveiby, K. (1989) *The Invisible Balance Sheet.*Stockholm: Affarfgarblen. 5

Sveiby, K.. (1997) *The New Organizational Wealth*. Stockholm: Berrett-Koehler Publishers. 5

Sveiby, K. (2001) A Knowledge-Based Theory of the Firm to Guide Strategy Formulation. *Journal of Intellectual* Capital, 2(4), 344–358. DOI: 10.1108/14691930110409651 5

Talero, E. and Gaudette, P. (1995) *Harnessing Information for Development: World Bank Group Vision and Strategy*. Washington, DC: World Bank.

TFPL. (2001) *Knowledge Strategies – Corporate Strategies*. TFPL's 4th International CKO Summit. London: TFPL.

Tsoukas, H. (1996) The firm as a distributed knowledge system: a constructionist approach. *Strategic Management Journal*, 17, 11–25. 21

Tushman, M. (1977) Special Boundary Roles in the Innovation Process. *Administrative Science Quarterly*, 22(4), 587–606. DOI: 10.2307/2392402 24

Tushman, M. and Scanlan, T. (1981a) *Characteristics and External Orientations of Boundary Spanning Individuals, Academy of Managemenr Journal*, 24(2), 83–98. DOI: 10.2307/255825 23, 24

Tushman, M. and Scanlan, T. (1981b) Boundary Spanning Individuals: Their Role in Information Transfer and Their Antecedents. *Academy of Management Journal*, 24(2), 289–305. DOI: 10.2307/255842 23, 24

von Krogh, G., Roos, J., and Kleine, M. (1998) *Knowing in firms: Understanding, managing and measuring knowledge*. London: Sage. 20

WebMD: Better Information, Better Health. `http://www.webmd.com/default.htm`

Weick, K. (1995) *Sensemaking in organizations*. Thousand Oaks, CA: Sage. 20

Wenger, E. (1998a) *Communities of practice: Learning, meaning and identity*. Cambridge: Cambridge University Press. 43

Wenger, E. (1998b) Communities of practice: Learning as a social system. *Systems Thinker*, 9(5), 1–5.

Wenger, E., and Snyder, W. M. (1999) Communities of practice: The organizational frontier. *Harvard Business Review*, 78(1), 139–145. 27, 43

Wilensky, H. (1967) *Organizational Intelligence: Knowledge and Policy in Government and Industry*. New York, NY: Basic Books.

Wilson, T. (2002) The nonsense of 'knowledge management', *Information Research* 8(1), Paper No. 144. 3, 45

Winograd, T. (1988) A language/action perspective on the design of cooperative work. *Human Computer Interaction*, 3(1), 3–30. DOI: 10.1207/s15327051hci0301_2 29

Yu, T., Lu, L., and Liu, T. (2009) Exploring factors that influence knowledge sharing behavior via weblogs. *Computers in Human Behavior*. DOI: 10.1016/j.chb.2009.08.002 49

Zack, M. (1999) Managing codified knowledge. *Sloan Management Review*, 40(4), 45–58. 35

Zhang, Y., Fang, Y., Wei, K., and Chen, H. (2010) Exploring the role of psychological safety in promoting the intention to continue sharing knowledge in virtual communities. *International Journal of Information Management, in press*. DOI: 10.1016/j.ijinfomgt.2010.02.003 49

Authors' Biographies

CLAIRE R. MCINERNEY

Claire R. McInerney is an Associate Professor and Chair of the Department of Library and Information Science in the School of Communication and Information at Rutgers, the State University of New Jersey. Her PhD in Information Science was earned from the State University of New York/Albany. She serves on the editorial board of the Journal of the American Society for Information Science and Technology (JASIST) and the Journal of Information and Knowledge Management. Prof. McInerney's research focuses on Community Informatics, Knowledge Management, and Health Informatics. She is the author of "Providing Data, Information and Knowledge to the Virtual Office" and is co-editor of "Re-Thinking Knowledge Management: From Knowledge Objects to Knowledge Processes." She teaches courses on Knowledge Management and is the director of the Knowledge Institute, a research center at Rutgers University.

MICHAEL E.D. KOENIG

Michael E.D. Koenig is Professor at Long Island University, and he is the former and founding Dean of the College of Information and Computer Science at LIU. His career has included both academic positions and senior management positions in the information industry, including Manager of Information Services for Pfizer Research, and V.P. positions at the Institute for Scientific Information, Swets and Zeitlinger, and Tradenet Inc., and Dean and Professor at Dominican University. Koenig obtained his Ph.D. in information science from Drexel University, an MBA in Mathematical Methods and Computers and an M.S. in Library and Information Science from the University of Chicago, and his undergraduate degree in Psychology and Physics is from Yale University. A Fulbright Scholar in Argentina, he is the author of more than 100 peer reviewed scholarly publications, the co-editor of three monographs on the subject of KM published for the American Society for Information Science and Technology, including *Knowledge Management: What Works and What Doesn't*, a member of the editorial board of more than a dozen journals, and a past president of the International Society for Scientometrics and Informetrics. He is also the recipient of a Jason Farradane award for "Outstanding Work in the Information Field."